Cortical Sensory Organization

Multiple Visual Areas

Cortical Sensory Organization

Edited by **Clinton N. Woolsey**

Volume 1: **Multiple Somatic Areas**
Volume 2: **Multiple Visual Areas**
Volume 3: **Multiple Auditory Areas**

Cortical Sensory Organization

Volume 2

Multiple Visual Areas

Edited by

Clinton N. Woolsey

University of Wisconsin, Madison, Wisconsin

Humana Press • **Clifton, New Jersey**

Library of Congress Cataloging in Publication Date

Main entry under title:

Cortical sensory organization.

  Includes bibliographical references and index.
  Contents: v. 1. Multiple somatic areas — v. 2. Mul-
tiple visual areas.
  1. Cerebral cortex.   2. Senses and sensation.
I. Woolsey, Clinton N.
QP383.C67   599.8′04182   81-81433
ISBN 0-89603-031-8 (v. 2)
AACR2

© 1981 The HUMANA Press Inc.
Crescent Manor
P. O. Box 2148
Clifton, N. J. 07015

ISBN 0-89603-031-8

Printed in the United States of America

Table of Contents

Chapter 2

Chapter 3

Chapter 4
Families of Related Cortical Areas in the Extrastriate Visual System: Summary of an Hypothesis

Ann M. Graybiel and **David M. Berson**

Chapter 5
Cortical and Subcortical Connections of Visual Cortex in Primates

Rosalyn E. Weller and **Jon H. Kaas**

Chapter 6
Organization of Extrastriate Visual Areas in the Macaque Monkey............................   *157*
D. C. Van Essen, J. H. R. Maunsell and J. L. Bixby

Chapter 7
Visual Topography and Function: Cortical Visual Areas in the Owl Monkey......................   *171*
John M. Allman, James F. Baker, William T. Newsome and Steven E. Petersen

Chapter 8
Cortical Visual Areas of the Temporal Lobe:
Three Areas in the Macaque *187*

**C. G. Gross, C. J. Bruce, R. Desimone,
J. Fleming** and **R. Gattass**

Contents of Other Volumes

Cortical Sensory Organization

Edited by **Clinton N. Woolsey**

Volume 1: Multiple Somatic Areas

Volume 3:   **Multiple Auditory Areas**

Preface

In April 1979 a symposium on *"Multiple Somatic Sensory Motor, Visual and Auditory Areas and Their Connectivities"* was held at the FASEB meeting in Dallas, Texas under the auspices of the Committee on the Nervous System of the American Physiological Society. The papers presented at that symposium are the basis of most of the substantially augmented, updated chapters in the three volumes of *Cortical Sensory Organization*. Only material in chapter 8 of volume 3 was not presented at that meeting.

The aim of the symposium was to review the present status of the field of cortical representation in the somatosensory, visual and auditory systems. Since the early 1940s, the number of recognized cortical areas related to each of these systems has been increasing until at present the number of visually related areas exceeds a dozen. Although the number is less for the somatic and auditory systems, these also are more numerous than they were earlier and are likely to increase still further since we may expect each system to have essentially the same number of areas related to it.

Discovery of second somatic, visual and auditory areas in the early 1940s followed soon after the development of the evoked potential method for the study of cortical localization. The great increase in the number of recognized areas in the last 10 years has resulted from the use of microelectrode recordings from small clusters of neurons and from single units, which permit far more detailed examination of the brain than did the technology of earlier years. Other factors have been the study of more lightly anesthetized animals and of unanesthetized animals, whose brains have been explored through chambers implanted over the areas to be studied. One can expect the number of recognized areas to increase as more of the cortical surface is explored in detail in various species of animals.

Most individual studies to date have dealt with a single system for which the investigators have developed specialized equipment and skills in its study. There is evidence, however, that some cortical areas may respond to more than one modality of sensory input. This is particularly true of the so-called "association" cortex of the suprasylvian gyrus of the cat. It now seems very desirable to explore under optimal conditions all cortical areas with stimuli of more than one sensory modality. This will require that investigators acquire sophisticated equipment for the study of somatic, visual and auditory systems and develop skills in its use. An alternative method would be for experts on each system to join forces, so that the methods specialized for the three systems can be applied in a single given experiment.

Increasing quantities of information on the organization of afferent and efferent systems are being derived from the application of techniques for the study of connectivities within the central nervous system, through the use of tritiated amino acids and horseradish peroxidase, as illustrated by several of the studies reported in these volumes.

An important area of research not covered in these volumes is the study of behaving animals with implanted recording electrodes. I foresee that ultimately all areas of the cortex will be examined in this way.

Another area requiring study is the sensory input to the cortical motor areas. Corticocortical connections to these areas have been studied, as reported in volume 1, but more detailed sensory input using electrophysiological methods have not yet defined the sensory inputs to the precentral and supplementary motor areas. Similarly, less work has been done on the motor output from the postcentral sensory areas and its relation to the sensory input to these areas. There is practically no modern work on the effects of electrical stimulation of the visual and auditory areas of the cortex, although motor effects were obtained on stimulation of these areas by Ferrier and other early students of cortical localization.

An important problem for the future concerns the terminology to be applied to the many new cortical areas. If these areas correspond to recognized cytoarchitectural areas of Brodmann, that terminology should be applied. At present there is considerable confusion in terminology, perhaps best illustrated by the terms used to describe the various auditory areas in cat and monkey, where terms for the cat are related to the position of the areas in the auditory region. However, because the auditory region changes its orientation with evolution, the same terms used for cat cannot be used for the monkey. Perhaps when all areas have been identified and their corticortical connections and relations with subcortical structures have been fully defined, a more rational terminology can be proposed.

The three volumes of this work do not include reports from all the important workers in the fields surveyed. It was not possible in the time available to the symposium to include all those we should have liked to invite.

The editor wishes to express his deep appreciation to all those who took part in the Dallas symposium, and to thank them for the manuscripts which they prepared for these three volumes on *Cortical Sensory Organization*. He is also grateful to Drs. J. C. Coulter, J. H. Kaas and J. F. Brugge, who chaired the three programs. Finally, special thanks is due to Thomas Lanigan of the Humana Press for his interest in publishing this work and the care that he has devoted to seeing the material through the press.

The editor also wishes to thank Evadine Olson for several typing tasks that she performed in relation to his editorial functions.

Chapter 1

Multiple Cortical Visual Areas

Visual Field Topography
in the Cat

R. J. Tusa,[1]
L. A. Palmer[2] and A. C. Rosenquist[2]

[1]*Department of Neurology, The Johns Hopkins Hospital, Baltimore, Maryland and* [2]*Department of Anatomy, School of Medicine, University of Pennsylvania, Philadelphia, Pennsylvania*

1. Introduction

Localization of function has been a fundamental problem both historically and conceptually. The subject is of special interest to students of the cerebral cortex since this tissue is not structurally divisible into parts in any obvious way. At the turn of the century, anatomists were able to divide cerebral cortex into several discrete areas using special stains (5,6). This was based on subtle differences in cell size, cell density and fiber bundles. It was also realized, in many cases, that the primary inputs to these areas arose from

1

different afferent systems. Thus, for example, Brodmann's area 17 was recognized on the basis of its striking lamination and was known to be the recipient of a major projection of the retina by way of the lateral geniculate. Some appreciation of the functional significance of these areas later became apparent through the use of electrophysiological techniques. By restricting the stimulus to a limited portion of the receptor surface, it was possible to map the representation of that receptor surface on the cortex. The ultimate, now classical, accomplishments here were those of Woolsey and his collaborators (25). They described two important classes of cortical representations of sensory and motor fields. First, the sensory surfaces (or motoneuron pools) were completely and singly represented in certain "primary" areas corresponding in many cases to areas defined using cytoarchitectonics by earlier workers. Second, Woolsey described ancillary representations of these same sensory and motor fields that lay adjacent to the primary areas.

More recently, detailed and exhaustive application of single unit recording methods have revealed numerous previously unexpected areas of cortex in which sensory and motor fields are represented. Each topographic representation corresponds to a histologically defined area that receives a unique set of anatomical projections. The work of Allman and Kaas (3) on visual cortex of owl monkey serves as a good example. They have found that area 19 of owl monkey actually includes at least five separate representations of the visual hemifield. Subsequently, it has been found that these five areas may also be distinguished on the basis of their connections with the rest of the brain and on their cytoarchitecture.

We adopt as a working hypothesis the idea that these discrete representations of the visual field are functional units of visual cortex, i.e., that they each make some contributions to visuomotor behavior or perception distinct from those of the other areas. This is not a new idea; rather, it has formed the philosophic substrate for studies of localization of function since the hypothesis was first overtly tested by Bouillaud (4). Since then, however, much research has come to support this idea. In the case of the visual areas, for example, it now seems clear that each area differs from its neighbors in its cytoarchitecture, its connections with the rest of the brain and in the properties of single units recorded there. In some instances, there is further support from behavioral studies.

In this chapter, we summarize our electrophysiological studies on the retinotopic organization of the visual cortex in the cat. As has been reported for the owl monkey and other species, we have found several areas containing retinotopically organized representations of the visual field. Our primary objective here is to show how the details of the visual field topography found in each area supports the idea that these areas constitute distinct functionally meaningful units. By comparing certain properties of visual field topography, we will demonstrate the uniqueness of each area in ways that seem likely to reflect their contributions to visual behavior. Of especial importance here are the type of topographic transformations of the visual field onto the cortical surface in each area, the extent of the visual field represented in each area and the extent to which the central or peripheral portions of the visual field are emphasized in each area. Finally, we will speculate on the significance of differences in these aspects of visual topography found in each of the areas.

2. Techniques

The detailed methods used in these studies have been described in a previous report (20). The topographies of visual field representations were determined by relating the positions of receptive fields for single or small clusters of neurons to the locations of the corresponding recording sites in the cortex of 100 cats. Penetrations with tungsten microelectrodes were made systematically across the cortex in anesthetized, paralyzed preparations. Receptive field positions were determined by moving rectangles of various sizes, orientations and velocities across a 6-ft diameter hemispheric screen centered in front of the cat's right eye. Small electrolytic lesions were made at the beginning and end of every penetration and full histological reconstruction was done using both the Woelcke fiber stain and the cresyl violet cellular stain to locate the electrode penetrations and relate them to the cytoarchitectonic borders. A styrofoam model of the cat cerebrum was constructed from coronal sections taken from an animal with the most common sulcal pattern. The model was enlarged 20 times and was corrected for the shrinkage that accompanied the celloidin embedding technique. Raw data were fitted onto this model as each experimental brain was reconstructed. Summary diagrams were taken directly from this model and appear in the latter pages of this chapter.

3. Visual Field Topography

3.1. Location of Cortical Visual Areas

The location of the visual field representations in cat cortex is illustrated in Fig. 1.1. One portion of cat cortex is divisible into five well-defined representations of the visual field. These include a single representation found in each of the histologically defined areas 17, 18 and 19 and two representations found in area 20 of Heath and Jones (8). The area of cortex within the suprasylvian sulcus and the posterior portion of the middle suprasylvian sulcus is

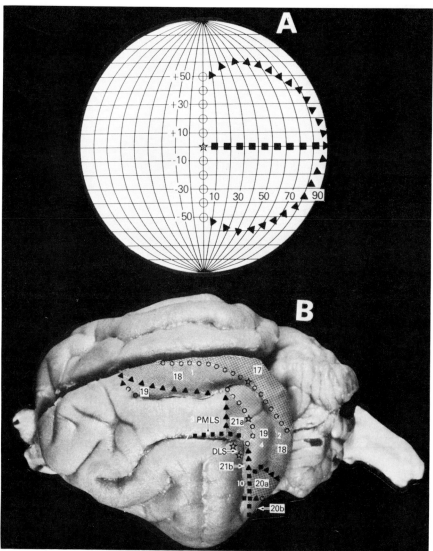

FIG. 1.1. Location of the 13 representations of the visual field on the cat's left hemisphere. A, perimeter chart of the cat's right visual hemifield. Various parts of the perimeter chart are illustrated with symbols, which are also appropriately placed onto the four views of the cat's left hemisphere (B–E) to show the location of the 13 visual areas. B, a dorsolateral view.

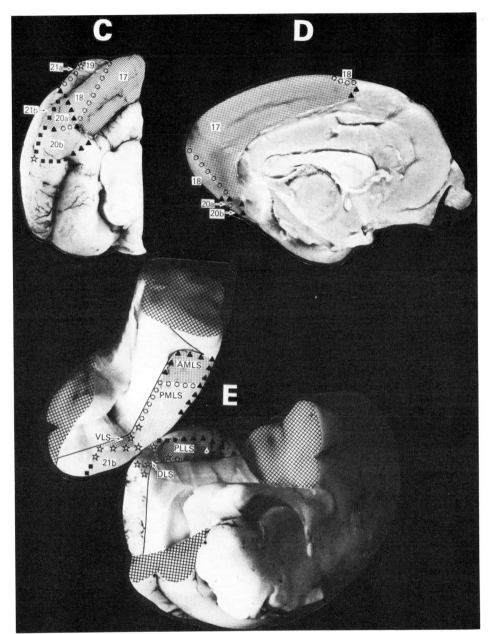

FIG.   1.1. (*continued*) C, posterior view. D, medial view. E, posteromedial view with the lateral [1], posterolateral [2], middle suprasylvian [3]
and posterior suprasylvian [4] gyri reflected up to expose the medial [5] and
lateral [6] banks of the middle suprasylvian sulcus [7] and the anterior [8]
and posterior [9] banks of the posterior suprasylvian sulcus [10]. From ref.
22.

very complex. On the basis of anatomical and physiological information presently available, this region appears to be composed of eight separate retinotopically organized areas (9, 11, 12, 15, 16, 18, 21–24). Six of these areas lie in Heath and Jones' (8) lateral suprasylvian area, three on the lateral bank of the middle and posterior suprasylvian sulci and three on the medial bank of these two sulci. Two other areas lie in Heath and Jones' (8) area 21.

3.2. *Visual Field Transformations*

The visual field represented in each of the cortical visual areas is not always in the form of a simple perimeter chart. There are basically four types of representations that are schematized in Fig. 1.2. The first type of transformation is a "point-to-point, first-order" transformation of the visual field (Fig. 1.2A). In this type, there is a point-to-point transformation of the visual field from the perimeter chart onto the cortex. In addition, all adjacent points in the perimeter chart are also adjacent in the cortical representation. The only distortion of the perimeter chart, as it is represented in the cortex, is that it is twisted and portions of it are magnified. A flattened map of the cat visual cortex showing the visual field representations is shown in Fig. 1.3. In this figure, AMLS best illustrates a point-to-point, first-order transformation of the visual field. Other areas containing this type of transformation include 17, 20a and ALLS.

The second type of transformation is the "point-to-point, second-order" transformation of the visual field (Fig. 1.2B). This type also contains a point-to-point transformation of the visual field from the perimeter chart onto cortex, but some adjacent points along the horizontal meridian in the visual field are not represented as adjacent points in the cortex. The perimeter chart, as it is represented in the cortex, must be split along the horizontal meridian so that this meridian is represented twice in widely separate cortical loci. Areas 18 and 19 both contain this type of transformation (Fig. 1.3).

A third type of transformation is a "point-to-line" transformation (Fig. 1.2C). There is no systematic representation of elevation in this type; instead, it is as if each point on the horizontal meridian is represented as a line of cortex. Areas VLS and DLS both contain this type of transformation (Fig. 1.3).

The last type of transformation is called a "mixed" transformation (Fig. 1.2E). This type of transformation contains a mixture of "point-to-point" and "point-to-line" transformations. Areas 20b, 21a, 21b, PMLS and PLLS all contain this type of transformation (Fig. 1.3).

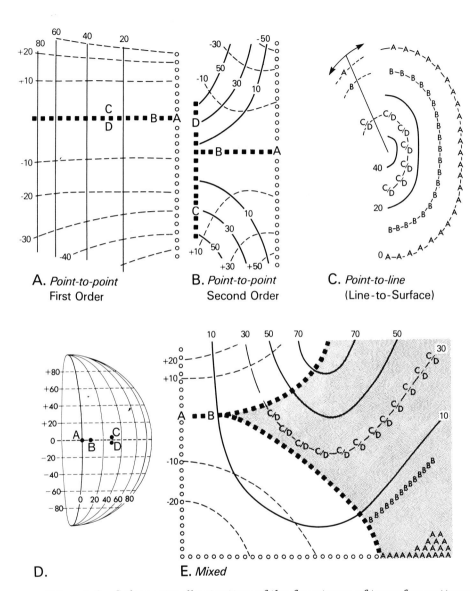

A. *Point-to-point*
First Order

B. *Point-to-point*
Second Order

C. *Point-to-line*
(Line-to-Surface)

D.

E. *Mixed*

FIG. 1.2. Schematic illustration of the four types of transformations of the visual field found in cat cortex. Four points on the perimeter chart (D) are labeled and their relative locations in the four types of visual field transformations are shown in Figs. A, B, C, and E. Point A represents the point of fixation, B represents the point 7° out along the horizontal meridian, C and D represent two points 30° out just above and below the horizontal meridian. From ref. 11.

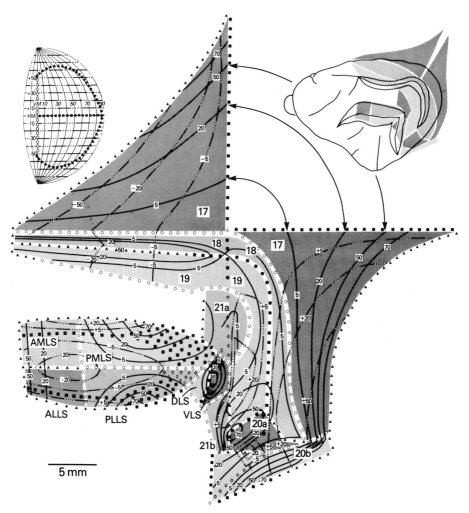

FIG. 1.3. A schematic unfolding of the visual cortex of the left hemisphere of the cat. The perimeter chart in the upper left corner shows the
contralateral (right) half of the visual field (as seen by the cat). The open diamond indicates the point of fixation, the circles indicate the vertical meridian, the solid squares the horizontal meridian, the solid triangles the
temporal periphery of the contralateral half of the visual field, the dashed
lines the isoelevation meridians and the solid lines the isoazimuth meridians. These symbols were then appropriately placed on the unfolded cortex,
illustrated in the center, to indicate the topographic organization of the
visual field within the 13 cortical areas. The cortex was unfolded from the
dorsolateral view of the cat's hemisphere in a manner similiar to the figure
in the upper right corner. The top of the figure represents the medial extent of the visual cortex, the bottom of the figure represents the lateral extent, the left part of the figure represents the rostral extent of cortex and
the right part of the figure represents the caudal extent of cortex. In order
to unfold the visual cortex, area 17 had to be separated along the horizontal meridian. From ref. 22.

8

The possible functional significances of point-to-point, first-order and second-order transformations of the visual field have been discussed by Allman and Kaas (2), and more recently by Allman (1). These two articles should be reviewed for a complete appreciation of their discussion. Reciprocal connections between homotypical points in visual space between adjacent areas would be much shorter, if one of these areas contained a first-order and the other contained a second-order transformation of the visual field than if they both contained the same type. The length of reciprocal connections between adjacent points just above and below the horizontal meridian in the area containing a second-order transformation would be greatly increased. These two facts suggest that areas containing second-order transformations of the visual field may be functional adjuncts to adjacent areas containing first-order transformations. As was discussed in an earlier paper (22), point-to-line transformations of the horizontal meridian, such as that seen in VLS and DLS, are the most efficient types of organizations for visual functions requiring extensive cellular interactions in the horizontal direction, such as stereopsis and control of vergent eye movements. The potential for cellular interactions between representations of adjacent points along a single isoelevation, such as the horizontal meridian, would be much more extensive in a point-to-line compared to a point-to-point transformation. The disadvantage of a point-to-line transformation is that cellular interactions between representations of adjacent points all found in a single isoazimuth, such as the vertical meridian, would be very restricted compared to a point-to-point transformation.

A mixed transformation has the advantages of point-to-point and point-to-line transformations, in that it can represent a fairly large portion of visual field and also increases the potential for extensive cellular interactions within certain portions of the visual field. For example, in areas 20b, 21a, and 21b, where the area centralis is represented as a line, the amount of cellular interactions for points in the central portion of the visual field would be increased compared to a strict point-to-point organization (Fig. 1.3).

3.3. *Extent of Visual Field Represented*

The portion of the visual field represented in each of the cortical areas differs (Fig. 1.4). Area 17 has the most complete representation of the visual field, probably as extensive as the visual field actually used by the behaving cat (Sprague, personal communication). Area 18 has only the central 50° of the binocular portion of the visual field represented. Area 19 has a limited representation of the vertical meridian, but has a complete representation of the hori-

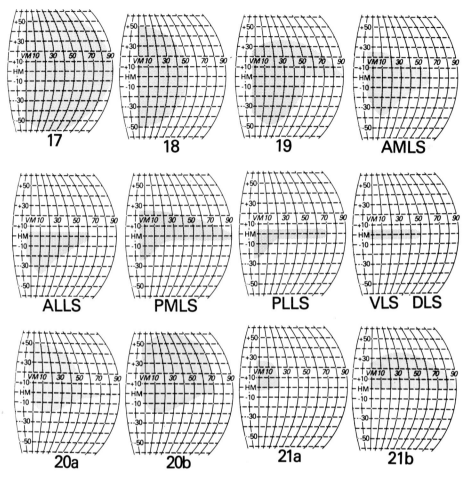

FIG. 1.4. Extent of the visual field represented in the 13 cortical visual areas.

zontal meridian. The four caudal areas within the suprasylvian sulcus (PMLS, PLLS, VLS, and DLS) all have an even more limited representation of the vertical meridian and an extensive representation of the whole horizontal meridian. The two rostral areas within the suprasylvian sulcus (AMLS and ALLS) have a larger representation of the lower fields than the upper fields, whereas areas 20a and 20b have a larger representation of the upper fields compared to the lower fields. Area 21a has less than the central 20° represented and area 21b has only the upper visual field represented. Presumably each area contains just sufficient visual field representation for the particular function(s) it performs. The elimination of excess visual field representation in each area effectively increases the magnitude of cortical surface area devoted to the portion of useful visual field representation.

3.4. Areal Magnification Factor

Magnification factor is a linear measurement of the magnitude of cortical surface area differentially devoted to various portions of the visual field. To illustrate the various magnification factors for the entire visual field representation, it is more accurate to calculate this factor as an areal measurement. This is because the amount of cortex devoted to one meridian is not necessarily identical to that devoted to the orthogonal meridian. This is particularly true in areas that are arranged as narrow belts, such as areas 18, 19, and 20b, since the isoelevation meridians are much shorter than the isoazimuth meridians (Fig. 1.2). Measurements of areal magnification factors (AMFs) are calculated by determining the cortical surface area devoted to every $10°$ by $10°$ block of visual field, each divided by 100. The units are, in this case, mm^2 of cortex/degree^2 of visual space. A three-dimensional graph illustrating the magnitude of cortical surface area differentially devoted to various portions of the visual field is generated by connecting the AMFs calculated for each area with smooth lines.

The AMFs of the visual field representation found in each of the 13 cortical areas differ (Fig. 1.5). The AMFs are greatest near the representation of the area centralis and decrease along the horizontal and vertical meridians in all areas except AMLS and ALLS (Fig. 1.5). In these two areas, AMF is greatest $15°$ out on the horizontal meridian and decreases along the horizontal meridian towards the periphery and the area centralis. The AMFs $5°$ out along the horizontal meridian ranges from a maximum of 0.84 mm^2/deg^2 in area 17 to a minimum of 0.006 mm^2/deg^2 in area ALLS.

On the basis of AMFs, the 13 cortical areas can be divided into those that appear to be specialized for either peripheral or central vision. Areas 20a, AMLS and ALLS appear to be specifically wired to emphasize the peripheral visual field and to deemphasize the central visual field. The peak AMFs of the area centralis representation in area 20a, AMLS and ALLS are all at least 1.5 orders of magnitude less than that found in area 17. On the other hand, the decrease in AMFs along the horizontal meridian is much less than the proportional decrease in retinal ganglion cell density along this meridian as determined from Stone (17) (Fig. 1.6A).

Area 17 and the remaining nine cortical areas, including 18, 19, 20b, 21a, 21b, PMLS, PLLS, VLS, and DLS, appear to be specialized for central vision. The AMFs of the area centralis representation in these nine areas are all within 1.5 orders of magnitude of that found for the area centralis representation of area 17. Areas 18, 19, 20b, 21a, 21b, PLLS, and DLS appear to be further specialized to deemphasize the periphery. The decrease in AMFs along the horizontal meridian in these areas is proportionally greater than the decrease in retinal ganglion cell density found along this merid-

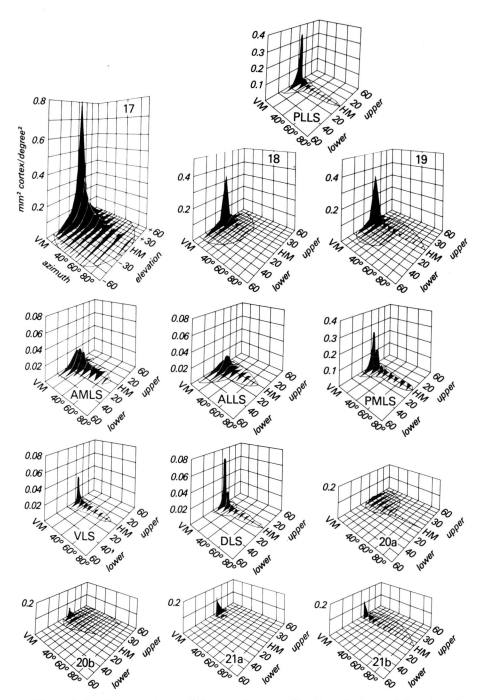

FIG. 1.5. Comparison of the areal magnification factors of the visual field represented in each of the 13 cortical areas in the cat. In each of the graphs, elevation is given on the abscissa and azimuth is given on the ordinate. Cortical surface area is given on the z axis. This measurement was calculated by determining the cortical surface area devoted to every 10° by 10° block of visual field. This point was then divided by 100 to give the point in mm² cortical surface area/degree² of visual field. The points along each azimuth were then joined by a line.

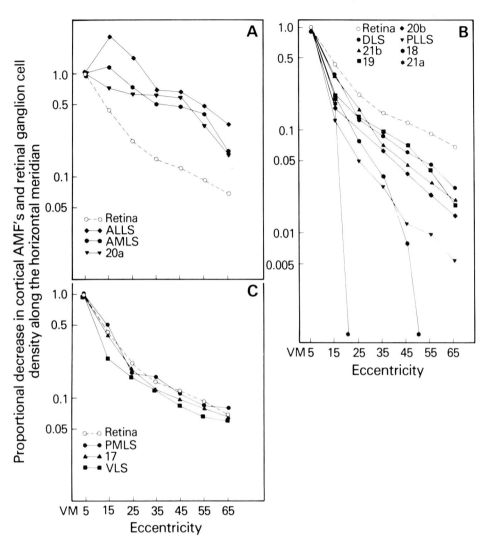

FIG. 1.6. Comparison of the areal magnification factors (AMFs) along the horizontal meridian and the retinal ganglion cell density along the horizontal meridian as determined by Stone (17) in the 13 cortical areas. For each area, the peak central AMF has been normalized with the peak central retinal ganglion cell density. A, illustrates those areas whose decrease in AMFs along the horizontal meridian is less than the decrease in retinal ganglion cell density. B, illustrates those areas whose decrease in AMFs along the horizontal meridian is equal to the decrease in retinal ganglion cell density. C, illustrates those areas whose decrease in AMFs along the horizontal meridian is more than the decrease in retinal ganglion cell density.

ian (Fig. 1.6B). These areas, compared to the retina, therefore, appear to be wired to deemphasize the peripheral fields. In this regard, areas 18 and 21a are so specialized that they just contain the central 50 degrees or less of the visual field. In areas 17, PMLS and VLS, the decreases in AMFs along the horizontal meridian are nearly the same as the proportional decrease in retinal ganglion cell density found along this meridian (Fig. 1.6C). Functionally, this suggests that these areas are no more specialized than the retina for processing information concerning the center of the visual field.

4. Unexpected Findings

4.1. Asymmetry

Striking asymmetries were found between the representations of the upper and lower visual fields within certain cortical areas (Fig. 1.4). The two rostral areas within the suprasylvian sulcus (AMLS and ALLS) have a larger representation of the lower fields than the upper fields, whereas 20a, 20b, 21a and 21b have a larger representation of the upper fields compared to the lower fields. There is also a pronounced asymmetry in the areal magnification factors in area 18 and especially area 19, where the peripheral 50 degrees of the lower visual field are not represented, although the corresponding portion of the visual field just above the horizontal meridian is represented. There are no obvious explanations for these asymmetries, but hopefully such explanations will arise as further work provides insight into the contributions these areas make to visual perception. One might speculate that as a result of the usual upright position of the cat near the ground, the types of visual stimuli presented to the upper and lower fields will differ. Perhaps certain areas are specialized to process these different types of stimuli. In this regard, it is interesting to note that there is a fairly extensive projection from areas 18 and 19 to the rostral pontine nuclei and that the projection from the representation of the lower visual hemifield is considerably denser than that from the representation of the upper visual field (14). The cells of origin of this projection respond best to full-field textured stimuli without preference to specific form, the majority of which prefer movement in the downward direction. Such cells are suitable for providing feedback about the terrain as an animal moves along the ground (7), further supporting the probable functional asymmetry between the upper and lower visual field representations in certain areas.

4.2. Variability

There is a considerable amount of variability found from cat to cat in the topographic organization of certain areas including area 17 (20). The degree of variability is best documented in the topographic organization of areas 18 and 19, where two basically different types of organizations are found among cats. For convenience, these two types have been designated as type 1 and type 2 organizations (21). A schematic representation of the visual hemifield in these two types on the unfolded cat cortex is shown in Fig. 1.7. Both types have the following characteristics. Area 18 (shaded) wraps around a portion of area 17 as a narrow belt. The border between areas 17 and 18 corresponds to a representation of the vertical meridian of the visual hemifield (open circles). The representation of the horizontal meridian (closed squares) in area 18 forms a "T" configuration, the base of which is the area centralis common to areas 17 and 18. This split representation of the horizontal meridian is shared with area 19 and, in fact, forms most of the boundary between areas 18 and 19. The remainder of the boundary separating the lower visual field representation between areas 18 and 19 consists of the representation of the far periphery. The representation of the visual hemifield within area 19 is basically mirror symmetrical with that in area 18.

In a type 1 organization, the representation of the isoelevation and isoazimuth lines constitutes a straightforward, mutually orthogonal lattice (Fig. 1.7B). In a type 2 organization, the representation of the isoazimuth lines are distorted and form two or more separate islands (Fig. 1.7C). At the center of these islands, receptive field centers extend out into the visual field as far as 50°, whereas, between these islands, receptive field centers are situated only 5° from the vertical meridian. There is a tremendous amount of variability among cats in the number and location of cortical islands in a type 2 organization (Fig. 1.8).

These islands and their variability from cat to cat create certain problems for anatomists. For example, a discrete injection of tritiated amino acids into an area projecting to area 18 would not necessarily result in a single terminal field, but rather in several separate patches corresponding to the location of these islands, which would vary from cat to cat. This was nicely demonstrated in a previous paper before the existence of these islands was known (13). A small injection of tritated leucine limited to the lateral

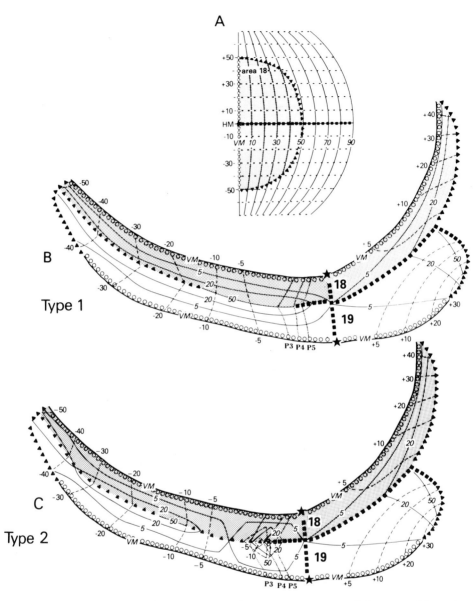

FIG. 1.7. Two different types of representations of the visual hemifield, labeled types 1 and 2, found in areas 18 and 19 on the unfolded cat cortex. A, perimeter chart based on a world coordinate scheme. The symbols used here are similar to those used in Fig. 1.3, except that a closed star indicates the point of fixation. B, a schematic illustration of a type 1 topography found in areas 18 and 19. The figure is oriented so that the rostral and caudal edges of the cortical areas, as they lie in the brain, are located at the left and right edges of the figure. The lateral edge of the cortical area is located at the bottom of the figure and the medial edge of the cortical area is located at the top. Area 18 is shaded. C, a schematic illustration of a type 2 representation of the visual hemifield in area 18 and 19 on the unfolded cat cortex.

16

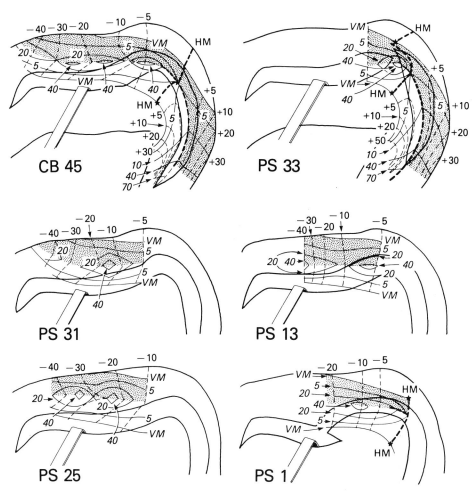

FIG. 1.8. Variability within the type 2 representation found in areas 18 and 19. This figure illustrates the dorsolateral views of six cat brains in which all or limited portions of areas 18 and 19 were mapped. The lateral and posterolateral sulci have been opened up. Area 18 is shaded and the representation of the horizontal meridian is symbolized by heavy dashed lines. Area centralis is symbolized by solid stars. From ref. 21.

portion of lamina A of the Dorsal Lateral Geniculate Nucleus (and to the Ventral Lateral Geniculate Nucleus which does not project to cortex) revealed a projection to the peripheral fields of areas 17 and 18 (Fig. 1.9). The extent of this projection is consistent with a region encompassing a visual field representation from 30° to 70° out in the field and from 5° to 30° below the horizontal meridian. A continuous streak of label was found in area 17, where no islands occur, but two separate patches of label were found in area 18, where they usually do occur.

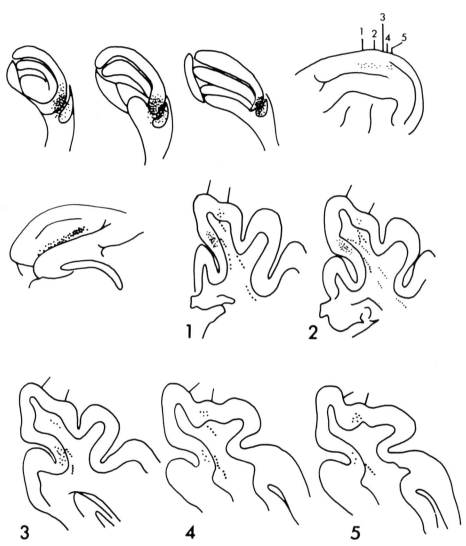

Fɪɢ.   1.9. Projections from the dorsal lateral geniculate nucleus to areas 17 and 18 in cat demonstrated by orthograde transport of tritiated amino acids (13). The injection site includes the monocular segment portion of lamina A and the ventral lateral geniculate nucleus, which does not project to the cortex. Coronal sections demonstrate the presence of label in areas 17 and 18. Note the appearance of label in area 18 at more rostral and caudal levels but not at intermediate levels. Dorsolateral and medial views of the hemisphere are shown primarily to permit visualization of the level of each coronal section. From ref. 13.

4.3. Ability of Topographic Representations of the Visual Field to Define Functional Areas

Parceling visual cortex on the basis of electrophysiological mapping techniques is based on the premise that the existence of a separate representation of the visual field, or some delimited portion of it, reflects a separate functional area. As was mentioned earlier in the chapter, five well-defined representations of the visual field are found in cat cortex. The areas these visual field representations coincided with have been labeled 17, 18, 19, 20a, and 20b. The visual field representations found in the suprasylvian sulcus and posterior portion of the middle suprasylvian gyrus, however, are much more complex. This complicated region has been divided into eight areas based on a number of techniques, including electrophysiological mapping, anatomical projections and cytoarchitecture. Based on topography alone, however, there are other ways of dividing this portion of cortex (11, 22). Further anatomical, physiological and behavioral studies of this region of cortex will either corroborate this particular division of cortex or suggest modifications.

5. Conclusion

We have demonstrated numerous topographic representations of the visual field in the cerebral cortex of the cat using microelectrophysiological mapping techniques. Each area containing a topographic representation appears to constitute a homogeneous unit based on its anatomical connections, single unit properties and cytoarchitecture. We suggest that each topographic representation constitutes a functional unit of cortex that makes as yet unspecified contributions to visuomotor behavior and perception. This concept is supported by the fact that each representation is unique in ways that seem likely to reflect its contributions to visual behavior. Specifically, we have described how these topographic representations differ with respect to the nature of transformation of the perimeter chart as found in the cortex, the portion of the visual field it contains, and the extent to which central or peripheral portions of the visual field are emphasized with respect to the visual field magnification factor.

These features of visual field representations can be used to suggest possible functions for certain areas. For example, vergent eye movements used to fixate objects in visual space at various distances from the organism appear to require two processes, initiation and fusion (10). Initiation of vergent eye movements is elicited by objects lying anywhere in the binocular field, but displaced from the horopter. This would require an area containing the entire binocular overlap zone, which is found only in areas 17 and 18. Since the velocities of saccades are much faster than vergent eye movements, completion of vergent eye movements (fusion) occurs with the object of interest now in the center of gaze (26). Fusion, therefore, does not require the entire binocular field, but does require an extensive analysis of retinal disparities, especially parallel to the horizontal meridian. Since areas with point-to-line representations of the horizontal meridian provide the maximal potential for analysis of such disparities, these areas are best suited to guide vergent eye movements to completion. PMLS, PLLS, VLS and DLS all contain this type of representation. On the other hand, afferent information used to generate saccadic eye movements would best be served by a visual field representation emphasizing the peripheral portion of the visual field such as that found in AMLS, ALLS and 20a.

In order to determine conclusively whether visual field representations constitute functional units, neural circuits underlying well-defined visual functions, such as vergent eye movements, must be determined. Pursuit of this type of research strategy will be facilitated now that visual field representations have been delineated in cat cortex.

6. Summary

The retinotopic organization of area 17 and visually responsive cortex surrounding this area has been recently reexamined in cat (11,20–22). Using electrophysiological mapping techniques, several representations of the visual hemifield have been found in cat cortex. Many of these representations correspond to cortical areas previously defined histologically and hodologically. Areas 17, 18 and 19 each contain a single representation, while the region of cortex in the vicinity of Heath and Jones' (8) area 20 contains two separate representations. The visual field topography found in the cortical regions labeled by Heath and Jones (8) as area 21 and the lateral suprasylvian area is very complicated. Based on anatomical and physiological information presently available, this region can be divided into eight areas. The location of these visual areas were located on a styrofoam model of the cat cerebrum constructed from coronal sections. The dorsolateral and medial views from this model are illustrated in Fig. 1.10. The individual sections, which were used to make the model, appear in Fig. 1.11, and the progression of the receptive field centers recorded in each coronal section is illustrated in Fig. 1.12.

The visual hemifield represented within these areas differs strikingly in a number of ways, perhaps corresponding to the function they perform in the visual system. Three of the more significant differences are described in this chapter. First, the topographic organization of the visual hemifield represented in these areas differs. In certain areas, the only distortion of the perimeter chart, as it is represented in the cortex, is the magnitude of cortical surface area differentially devoted to various portions of the visual hemifield. In other areas, the perimeter chart, as it is represented in the cortex, must be split along the horizontal meridian so that this meridian is represented twice in widely separate cortical loci. In other areas, there is no systematic representation of elevation; instead, it is as if each point on the horizontal meridian is represented as a line of cortex. The significance of these types of transformations is discussed. Second, these areas differ with respect to the portion of the visual hemifield represented. Area 17 is the only area containing a complete representation of the visual hemifield. All other areas contain representations of only certain portions of the visual hemifeld, such as the binocular portion of the visual field found in area 18 and the horizontal meridian found in some of the lateral suprasylvian areas. Third, the magnification factor of the portions of the visual field represented in these areas differs. Areas appear to be specifically wired to emphasize either the central or the peripheral visual field.

Finally, three unusual findings concerning visual field topography in cat are discussed. First, striking asymmetries were found between the representations of the upper and lower visual hemifields within certain areas among cats. Second, there is a certain degree of variability in the topographic organization of the visual hemifield found in these areas. Third, the internal organizations of certain regions are so complex that their subdivision must rely on additional techniques.

Acknowledgments

We thank the *Journal of Comparative Neurology* for allowing us to reproduce Figs. 1–3, 8, and 9. We also thank Doctor Susan J. Herdman for her helpful comments and Mr. John Woolsey and Mrs. Betty Woolsey for assistance with the illustrations.

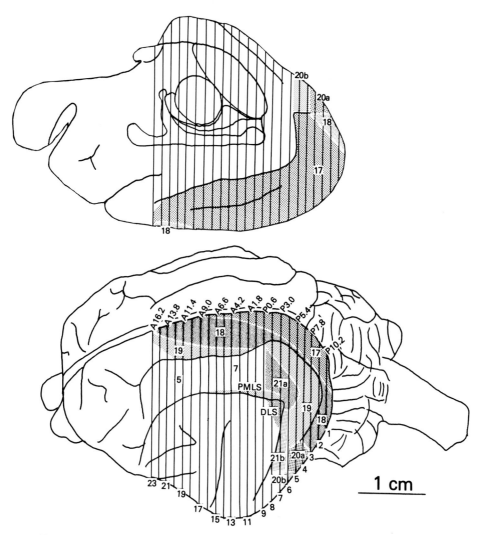

FIG.   1.10. Dorsolateral and medial views of the left hemisphere in
the cat showing the location of the cortical visual areas determined electrophysiologically. This figure also shows the location of coronal sections,
numbered at the bottom, which are illustrated in Fig. 1.11. The Horsley-
Clarke number of each coronal section is given at the top of the dorsolateral view.

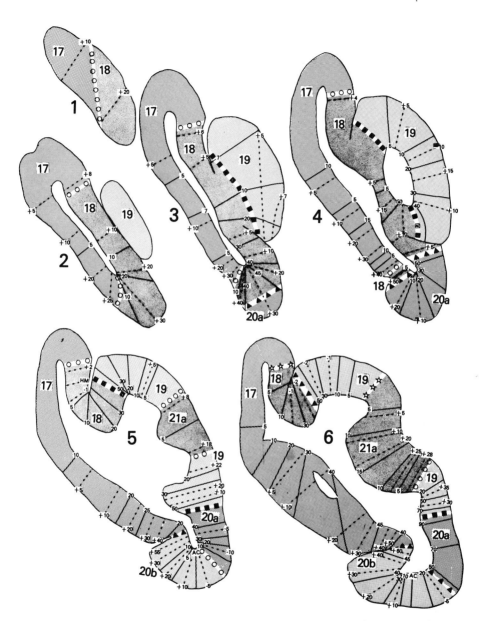

FIG. 1.11. Twenty-three coronal sections spaced every 1.2 mm illustrating the retinotopic organization of the visual field typically found within them. Sufficient variability in the retinotopic organization exists among cats that an exact map of the areas cannot be used for all cats. The purpose of the maps provided in these figures is to show the general topography of the areas.

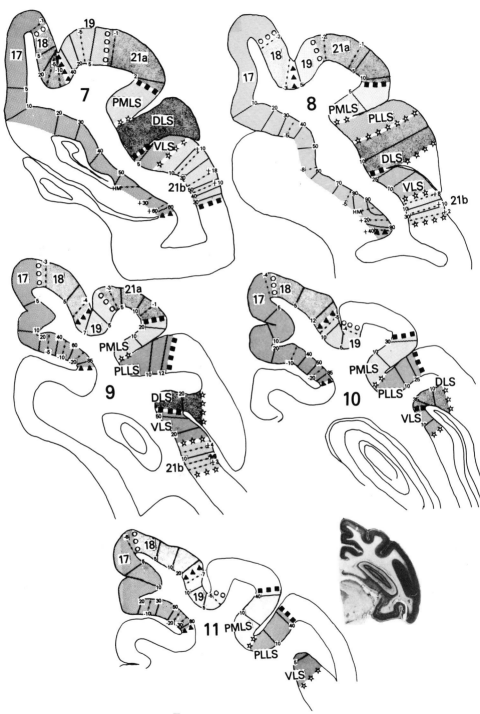

FIG.   1.11 *(continued)*

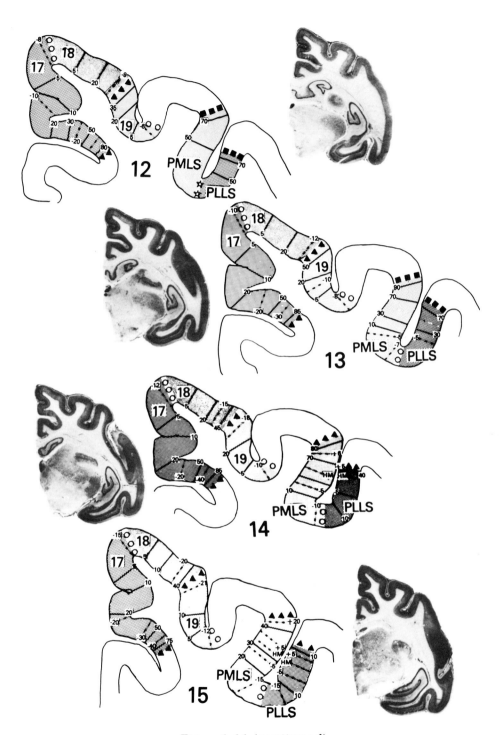

FIG. 1.11 (*continued*)

25

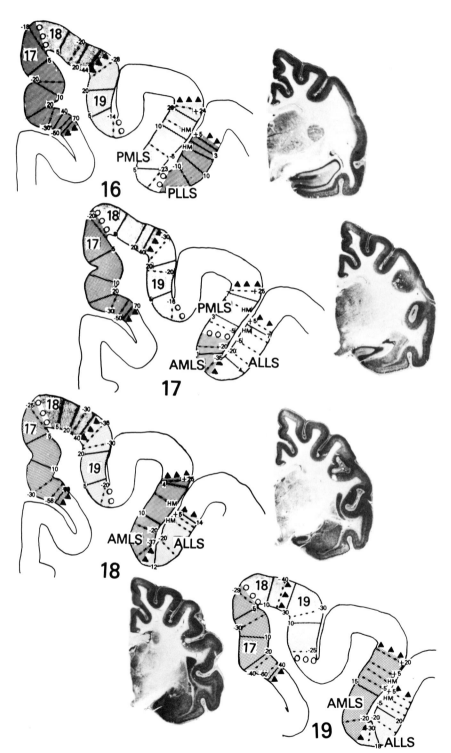

FIG. 1.11 (continued)

26

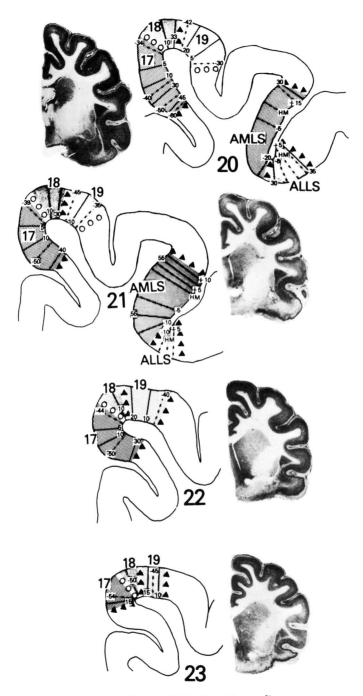

FIG. 1.11 *(continued)*

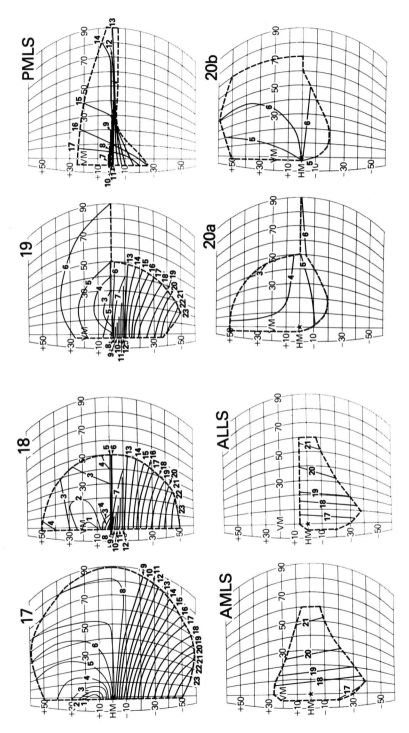

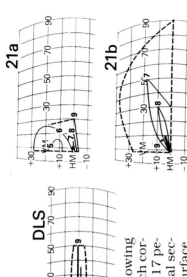

FIG. 1.12. Perimeter charts for each of the 13 cortical areas showing the progression of the centers of receptive fields recorded within each coronal section illustrated in Fig. 1.11. Lines 1, 2, 3 . . .23 in the area 17 perimeter chart trace out the distribution of receptive fields in coronal sections in Fig. 1.11, starting at the 17/18 border on the ventral surface. Similarly, each line in the perimeter chart in the upper center of Fig. 1.12 traces out the distribution of receptive fields in that portion of area 18, which lies in coronal sections numbered 1–23 illustrated in Fig. 1.11. This area is more complicated, however, in that portions of it are intercalated between other areas in some of the coronal sections. For example, line 3 in the area 18 perimeter chart (Fig. 1.12) is divided into three parts. The line beginning at 6° up on the vertical meridian and ending 7° out along the horizontal meridian corresponds to the portion of area 18 beginning at the 17/18 border on the dorsal surface of coronal section number 3 (Fig. 1.11) and ending at the 18/19 border in the fundus of the lateral sulcus. The line 3 in the area 18 perimeter chart (Fig. 1.12), beginning at 30° out along the horizontal meridian and ending 25° up by 45° out in the periphery, corresponds to that portion of area 18 beginning at the 19/18 border in the fundus of the posterior lateral sulcus of coronal section number 3 (Fig. 1.11) and ending at the 18/20a border at the far ventral surface of the coronal section. The last segment of line 3 in the area 18 perimeter chart (Fig. 1.12) corresponds to that portion of area 18 found between the 20a/18 border and ends at the 18/17 border on the ventral tip of coronal section number 3 in Fig. 1.11.

References

1. ALLMAN, J. Evolution of the visual system in the early primates. In: *Psychobiology and Physiological Psychology*, vol. 7, edited by J. SPRAGUE AND A. EPSTEIN. New York: Academic Press, 1977.

2. ALLMAN, J. M., AND KAAS, J. H. The organization of the second visual area (VII) in the owl monkey: a second order transformation of the visual hemifield. *Brain Res.*, 76: 247–265, 1974.

3. ALLMAN, J. M., AND KAAS, J. H. The dorsomedial cortical visual area: a third tier area in the occipital lobe of the owl monkey (*Aotus trivirgatus*). *Brain Res.*, 100: 473–487, 1975.

4. BOUILLAUD, J. Cited in: POLYAK, S. *The Vertebrate Visual System*, edited by H. Klüver. Chicago: University of Chicago Press, 1957, pp. 138–143.

5. BRODMANN, K. *Vergleichende Lokalisationslehre der Grosshirnrinde.* Leipzig: Barth, 1909, 324 pp.

6. CAMPBELL, A. W. *Histological Studies on the Localization of Cerebral Function.* Cambridge: Cambridge University Press, 1905, 360 pp.

7. GIBSON, A., BAKER, J., MOWER, G., AND GLICKSTEIN, M. Corticopontine cells in area 18 of the cat. *J. Neurophysiol.*, 41: 484–495, 1978.

8. HEATH, C. J., AND JONES, E. G. The anatomical organization of the suprasylvian gyrus of the cat. *Ergebn. Anat. Entwickl.*, 45: 1–64, 1974.

9. HUGHES, H. C. Thalamic afferents to two visual areas in the lateral suprasylvian sulcus of the cat. *Anat. Rec.*, 190: 426–427, 1978.

10. MITCHELL, D. E. Properties of stimuli eliciting vergent eye movements and stereopsis. *Vision Res.*, 10: 145–162, 1970.

11. PALMER, L. A., ROSENQUIST, A. C., AND TUSA, R. J. The retinotopic organization of the lateral suprasylvian areas in the cat. *J. Comp. Neurol.*, 177: 237–256, 1978.

12. RACZKOWSKI, D., Connections of the lateral posterior-pulvinar complex with the extrastriate visual cortex in the cat. *Soc. Neurosci. Abst.*, 5: 803, 1979.

13. ROSENQUIST, A. C., EDWARDS, S. B., AND PALMER, L. A. An autoradiographic study of the projections of the dorsal lateral geniculate nucleus and the posterior nucleus in the cat. *Brain Res.*, 80: 71–93, 1974.

14. SANIDES, D., FRIES, W., AND ALBUS, K. The corticopontine projection from the visual cortex of the cat: An autoradiographic investigation. *J. Comp. Neurol.*, 179: 77–88, 1978.

15. SEGRAVES, M. A. Cortical afferents to two visual areas in the lateral suprasylvian sulcus of the cat. *Anat. Rec.*, 190: 57, 1978.

16. SEGRAVES, M. A. Interhemispheric connections of the retinotopically defined visual cortical areas in the cat. *Soc. Neurosci. Abst.*, 5: 807, 1979.

17. STONE, J. The number and distribution of ganglion cells in the cat's retina. *J. Comp. Neurol.*, 180: 753–773, 1978.

18. SYMONDS, L., ROSENQUIST, A. C., EDWARDS, S., AND PALMER, L. Thalamic projections to electrophysiologically defined visual areas in the cat. *Soc. Neurosci. Abst.*, 4: 647, 1978.

19. SYMONDS, L., AND ROSENQUIST, A. C. Visual cortical input to area 20 of the cat: Anatomical evidence for subdivision of this region into two areas. *Soc. Neurosci. Abst.*, 5: 809, 1979.

20. TUSA, R. J., PALMER, L. A., AND ROSENQUIST, A. C. The retinotopic organization of area 17 (striate cortex) in the cat. *J. Comp. Neurol.*, 177: 213–236, 1978.

21. TUSA, R. J., ROSENQUIST, A. C., AND PALMER, L. A. Retinotopic organization of areas 18 and 19 in the cat. *J. Comp. Neurol.*, 185: 657–678, 1979.

22. TUSA, R. J., AND PALMER, L. A. Retinotopic organization of areas 20 and 21 in the cat. *J. Comp. Neurol.*, 193: 147–164, 1980.

23. UPDYKE, B. V. Projections from lateral suprasylvian cortex to the lateral posterior complex in the cat. *Anat. Abst.*, 193: 707–708, 1979.

24. UPDYKE, B. V. Evidence for additional extrastriate cortical projections to the dorsal lateral geniculate nucleus in the cat. *Soc. Neurosci. Abst.*, 5: 812, 1979.

25. WOOLSEY, C. N. Cortical localization as defined by evoked potential and electrical stimulation methods. In: *Cerebral Localization and Organization.*, edited by G. SCHALTENBRAND AND C. N. WOOLSEY. Madison: University of Wisconsin Press, 1964.

26. YARBUS, A. L. Motion of the eye on interchanging fixation points at rest in space. *Biofizika*, 2: 698–702, 1957.

Chapter 2

Comparative Studies on the Visual Cortex

Vicente M. Montero

Department of Neurophysiology, Waisman Center, University of Wisconsin, Madison, Wisconsin

1. Introduction

This review focuses on studies about the organization of the visual cortex in the rat, rabbit and cat that my colleagues and I have carried out in the past years. The unifying theme is the definition of extrastriate visual areas based on the pattern of distribution of striate–extrastriate cortical connections and their comparison in these species. Relevant electrophysiological, anatomical and behavioral studies on the visual areas of these and other related mammalian species are also considered.

2. Extension and Retinotopy of the Striate and Extrastriate Visual Areas in the Rat

In 1964, Rojas et al. reported in an electrophysiological study (63) the existence of three retinotopically different extrastriate areas in the rat located anteromedially, anterolaterally and laterally to the

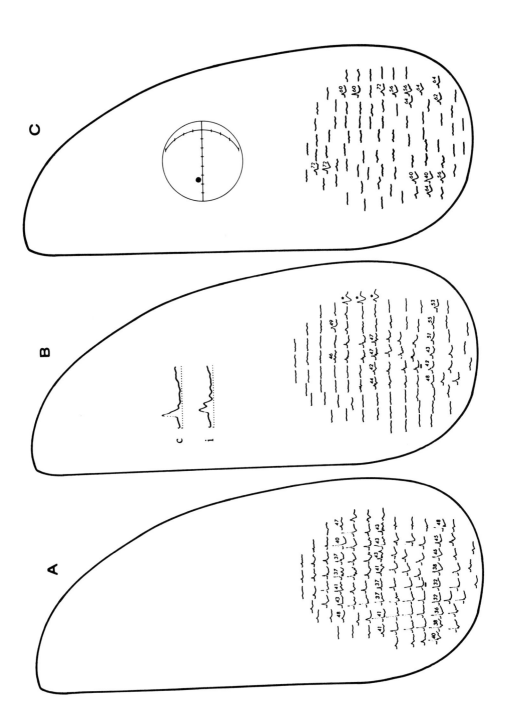

FIG. 2.1 Cortical responses to gross contralateral (A), ipsilateral (B), and to restricted (C) photic stimulation in the rat. The records are the average of the activity recorded within the cortex at each cortical point after 64 stimulus presentations. The numbers indicate the latency to the peak of the first negative, upward deflection. Potentials tagged by stars in B are acoustic responses (note their short latencies). B, upper insert: magnified pictures of selected contralateral (c) and ipsilateral (i) potentials. The potentials that are enlarged here are indicated by heavy underlining in A and B. Time scale subdivisions: 6 ms. C, upper insert: position of the 10° flash spot (black) in relation to the zero horizontal and zero vertical meridians of the contralateral visual field. Scale to the right in mm. Vertical calibration: 800 μV for A and B; 400 μV for C. Horizontal calibration: 150 ms for A and B; 200 ms for C (from Montero, 1973, ref. 39).

striate cortex. In the anteromedial area, the upper temporal visual field is rostral and medial and the inferonasal visual field is lateral and caudal. In the anterolateral area, the upper temporal visual field is rostrolateral and the inferonasal visual field is caudomedial. In the lateral area, the upper temporal visual field is caudomedial and the inferonasal visual field is rostromedial. In a subsequent study on the visual cortex of the rat, Adams and Forrester (1) mentioned a mirror-image secondary area lateral to the primary visual area and a region of undefined visual responses anterior to area 17.

In subsequent experiments (39), I decided to define, using microelectrode recordings of field potentials, the extent of the rat's occipital cortex responding to the gross photic stimulation of the contralateral and ipsilateral eye and also to determine whether multiple regions are activated in the visually responsive cortex by photic stimulation of a restricted part of the visual field (as would be expected by the previous retinotopic map).

Figure 2.1A and B show the results from a rat, in which the right posterior neocortical responses to contralateral and ipsilateral eye stimulation were mapped from 100 microelectrode penetrations. The contralateral field of responses extended from the occipital pole to 6.2 mm rostrally (about 37% of the hemisphere length) and its lateral and medial borders ran 5.8 and 1–2.5 mm, respectively, from the midline. The zone of ipsilateral responses (Fig. 2.1B) occupied the central region of the contralateral field, confirming previous results (32), and it showed lateral expansions posteriorly and anteriorly. This ipsilateral field, or zone of binocular vision, extends medially at least to the 50° vertical meridian projection in the primary visual area (50) at its wider part (see Fig. 2.2). The anterolateral and posterolateral expansions, showing potentials of increasing amplitudes, appear to correspond respectively to the anterolateral peristriate visual area (area AL, Fig. 2.2) and to the region, where a second reversal of the visual field is found posterolaterally (area LL, Fig. 2.2). The three short latency potentials (24–27 ms) tagged with a star laterally in Fig. 2.1B, and the corresponding ones in Fig. 2.1A, are responses from the acoustic cortex to the flashes' sound artifacts, thus revealing the neat boundary and lack of overlap between the visual and auditory cortex.

Figure 2.1C shows the different cortical zones activated by a single restricted stimulation in the periphery of the contralateral eye's visual field in a rat, whose posterior cortex was explored by 85 microelectrode penetrations. A 10° spot flash stimulation was projected on the campimeter at the intersection of 110° azimuth (angular distance from the vertical meridian) and + 10° elevation (angular distance from the horizontal meridian) of the visual field. The upper insert in Fig. 2.1C shows the position and size of the stimulus with respect to the zero horizontal and zero vertical meridians. The posteromedial group of responses represents the activated region in the primary visual area, since they match topographically the zone predicted by the retinotopic map (see Fig. 2.2). Furthermore, at the point where the shortest latency (40 ms) and largest amplitude potential was obtained, the receptive fields of clusters of units recorded in that penetration coincided quite precisely with the stimulus size and location. No evoked activity was recorded in the rest of the cortical extent of the primary visual area, i.e., up to 5 mm from the midline and 5 mm from the occipital pole (at 3 mm laterally). Beyond the silent region, however, evoked responses were elicited again, and with longer latencies than 40 ms, in restricted zones anteromedially (72 ms), anterolaterally (60 ms) and in a more extensive zone laterally (56–72 ms). These responses are by themselves direct evidence that the stimulated visual field sector is represented in multiple cortical regions outside the primary visual area. They lie within the limits of the whole posterior neocortex responding to the gross photic stimulation of the contralateral eye (compare Figs. 2.1A and 2.1C). By their cortical positions, the anteromedial and anterolateral groups of responses appear to be related to the corresponding cortical regions beyond the primary visual area, where a different retinotopic arrangement is found electrophysiologically (Fig. 2.2) and to which direct corticocortical connections from the striate area have been determined anatomically in the rat (see Fig. 2.9). However, the correlation of the lateral group of responses with the retinotopic and anatomical findings, which have shown an heterogeneous organization in this area, is not so readily evident. Only the anterior two-thirds in this responsive field, showing a gradient in the latencies and amplitudes of potentials anteriorly and posteriorly, can be reasonably equated to what has been identified retinotopically (Fig. 2.2) and anatomically (Fig. 2.9) as the lateromedial area.

The experiments described above demonstrated that most, if not all, of the visually responsive cortex in the rat lies in the dorsolateral aspects of the occipital cortex. This situation made possible a fairly complete retinotopic map of the striate and extrastriate visual cortex in this animal. This map is shown in Fig. 2.2. In this new study (50), three other extrastriate retinotopic areas were recognized in addition to the previously found three extrastriate areas. A standardized projection of the optic disc on the campimeter, at the intersection of the +15° isoelevation line and 75° isoazimuth line, was used to combine results from different experiments. Since in the retina of the rat there is a higher concentration of ganglion cells (central area) located 1 mm (20°) upward and tem-

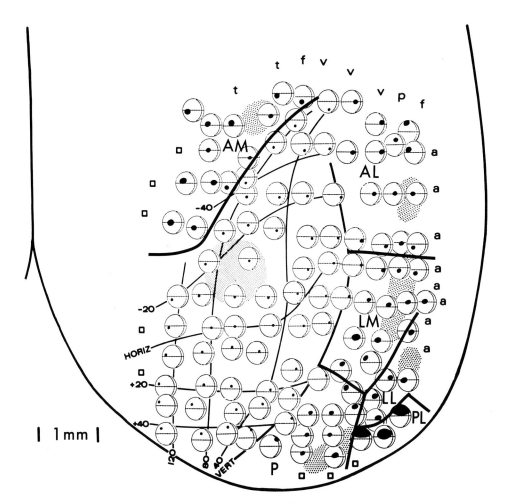

Fɪɢ. 2.2. Retinotopic organization of striate and extrastriate visual areas in the rat. In a dorsal view of the right posterior neocortex, the black regions in the field charts represent the receptive fields in the subjacent cortex in relation to the zero horizontal and zero vertical meridians of the left eye's visual field. The coordinates of the visual field representation in the striate cortex are indicated by isograde azimuth and elevation lines. The boundaries of the different extrastriate areas, in which there is a change in visual field directions, are indicated by thick lines: a, acoustic responses; t, f, v, p, somatic responses (see text); squares, nonresponsive penetrations. The shaded regions represent the approximate locations of a lesion in the striate cortex and the ensuing degeneration fields in the extrastriate cortex. (adapted from Montero, Rojas and Torrealba, 1973, ref. 50).

poral, in a diagonal direction, from the optic disc (16), the horizontal meridian of our map (15° below the optic disc) was equivalent to the true horizontal meridian of the rat's visual field.

The primary visual area presents a distinct and precisely arranged retinotopic organization, clearly observable in the progression of receptive fields throughout the region denoted by the isoelevation and isoazimuth coordinates in Fig. 2.2. The rostral and caudal regions within this area correspond to the inferior and superior visual fields, whereas the medial and lateral regions correspond to the temporal and nasal visual fields. This retinotopic pattern is analogous to that determined for the primary visual area in all mammals so far studied. The receptive field sizes in this area were characteristically the smallest determined (10–20°). Also, a proportionately larger expanse of cortex comprised the projection of central versus peripheral vision. Electrolytic lesions placed at the physiological boundaries of the primary visual area coincided with the limits of the densely packed granular cell layer IV that is characteristic of the striate cortex in the rat.

Beyond the primary visual area, visual responses characterized by large receptive fields (40–60°) were obtained in the lateral (18a) and anteromedial (anterior part of 18) peristriate cortices. The progressions of receptive fields along the mediolateral and rostrocaudal dimensions in these cortical regions not only differed from the trend observed in the primary area, but also showed different retinotopic arrangements between themselves. We have named the different extrastriate areas according to their positions relative to the striate cortex as follows (counterclockwise, in Fig. 2.2): area posterior (P), area posterolateral (PL), area laterolateral (LL), area lateromedial (LM), area anterolateral (AL) and area anteromedial (AM). The black spots on the field charts represent receptive fields of units recorded at subjacent cortical points with respect to the horizontal and vertical meridians. In areas AL, LM and P, the receptive fields move from nasal to temporal in a mediolateral direction. However, in a rostrocaudal direction, they change from up to down in area AL, from down to up in area LM and from up to down in area P. In area LL, there is a second nasotemporal reversal mediolaterally and lower and upper parts of the visual fields are rostral and caudal, respectively. In the area PL, in the posterior temporal cortex, the receptive fields are huge, precluding any retinotopy. In AM the receptive fields move towards the temporal periphery, medially. The completeness of the map is indicated by the limits of visual responsive cortex, with regions of no response (squares) medially and caudally, with acoustic responses laterally (a) and with somatic responses rostrally (t, trunk; f, forelimb; v, vibrissae; p, pinna).

The retinotopic subdivisions of several extrastriate visual areas in the rat are matched by corresponding pathways of striate–extrastriate connections, as is shown below. We have concluded from these studies that each of these extrastriate areas represents separate cortical functional units, that, among other possible functions, process in parallel information derived from striate cortex inputs. The nature of these processes remains to be determined.

3. Receptive Field Properties of Neurons in the Striate Cortex of the Rat

In collaboration with F. Torrealba and M.A. Carrasco (73), we performed a study of the receptive field properties of neurons of the striate cortex in gray rats, using extracellular recordings in paralyzed preparations. A brief description of the types of receptive fields found is given in this review.

The receptive fields of isolated neurons (N: 80) were classified as nonoriented (20%) and oriented (80%).

1. Nonoriented receptive fields were subdivided into (a) a concentric type (8%), with "on" and "off" centers and inhibitory periphery (these neurons responded to moving or stationary stimuli, provided they were of the appropriate size and contrast) and (b) a uniform type (12%) that discharged to "on", "off" or "on-off"of light everywhere in the receptive field. They had the largest receptive field size (10–50°). Some of these cells responded selectively to changes of general illumination. For others, the speed of a moving stimulus was critical.

2. Oriented receptive fields had either an orientation axis or a directional axis, or both. These receptive fields were classified as simple, complex and hypercomplex, according to the criteria of Hubel and Wiesel and according to their trigger features.

(a) A slit-type (20%) discharged optimally to a moving slit of light, within a restricted angular orientation, either directionally or nondirectionally. Elongating the slit beyond an optimal length did not affect the responses. Only three cells, that discharged to stationary slits or spots, could be classified as simple type.

(b) A dark bar-type had similar properties to the previous type, but responded to a dark bar. Only two cells fell into this category.

(c) An edge-type (20%) discharged to a moving or stationary edge of a specific contrast within a critical orientation (see Figs. 2.3–4). They could be directional or nondirectional. The response to the critical edge occured everywhere within the extension of the receptive field. All these cells could be classified as complex.

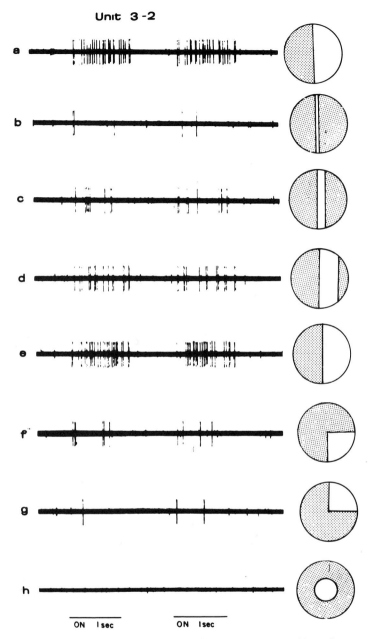

Unit 3-2

FIG. 2.3. Responses of an "edge" type complex cell in the striate cortex of the rat to several stationary stimulus configurations presented 1 s on, 1 s off in the receptive field. a and e, Responses to optimum stimulus, i.e., vertical edge, dark to the left. Note the sustained response during the stationary presentation of the stimulus. The unit responded to the optimal edge everywhere throughout an extension of 10° (not shown). d, Inhibitory effects elicited by the introduction of a border of opposite sign in the periphery of the receptive field. The inhibition increases in c and b. f and g, Show drastic diminution in the response by the intersection of 90° edge. h, No response to a 10° spot of light.

41

42

V. M. Montero

Unit 3-2

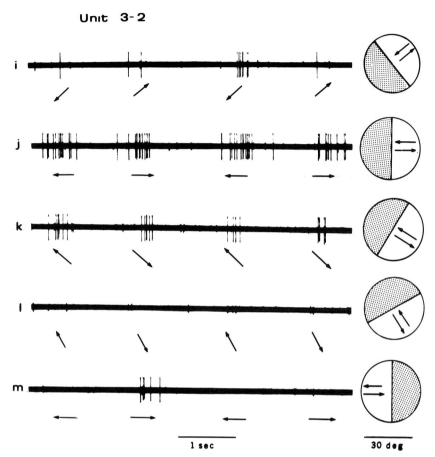

FIG. 2.4. Same "edge" type unit of previous figure, showing the responses to moving edges with different orientations. Note nondirectional responses to the optimum orientation in trace j. In traces i, k and l, responses at −45°, +30° and +60° from the vertical. m, No response to a vertical reciprocal edge.

(d) A contrast-type (8%) cell discharged to any linear boundary of contrast (slits, bars, edges) within a restricted orientation. They could be directional or nondirectional. All these cells could be classified as complex.

(e) A directional to small object type (18%) discharged to a small dark or light spot moving in a preferred direction (see Fig. 2.5). All these receptive fields were encountered in the upper half of the cortex. These cells could be classified in the hypercomplex category.

(f) A stopped-edge-type (10%) cell required for discharge strict angular configurations of a moving stimulus, directional (see Fig. 2.6) or nondirectional. These units also fit the hypercomplex category.

Unit 11-1

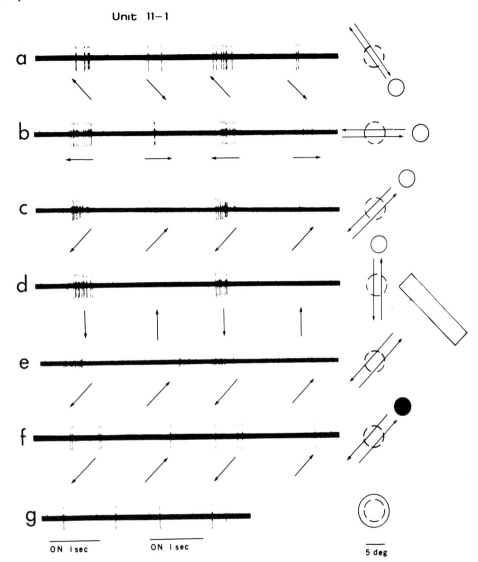

FIG. 2.5 Responses of a "directional to small bright object" hyper-complex type cell to moving stimuli of various configurations and directions. Note the wide angular range of the directional response in b, c and d, and the strong inhibition by a slit of light in e. To a stationary spot of light (g) the unit discharged one spike at on and at off.

The responses of a unique neuron are shown in Fig. 2.7. This unit had no spontaneous activity and responded poorly to dark bars receding into the receptive field with different inclinations (traces a–c). The unit, however, responded vigorously to a dark spot decreasing in diameter in the center of the field (trace d, right) and had no response to an expanding spot (trace d,left).

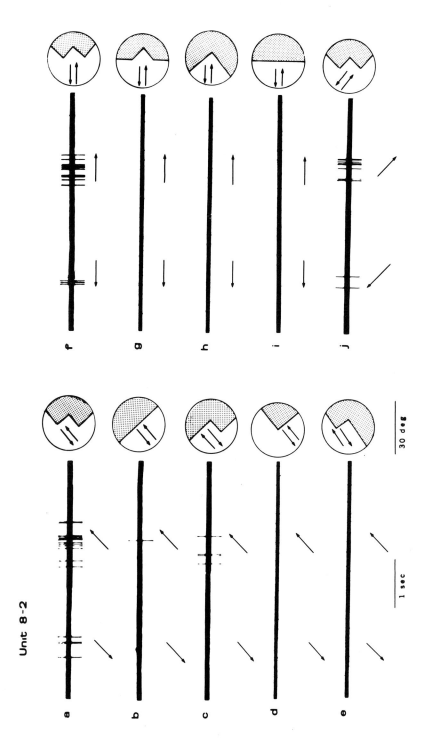

FIG. 2.6. Responses of a hypercomplex cell requiring a strict configuration of a double corner (a,f,j). Any modification of these geometrical configurations induced no response of the cell (b,c,d,e,g,h,i). Also, the response was directionally selective.

Unit I-2

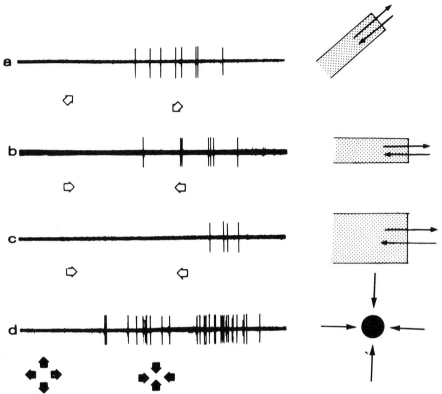

Fig. 2.7. Responses of a unit whose optimum trigger feature was a dark spot decreasing in diameter.

> This unit, "detector of a decreasing dark spot," would discharge in real life to a small dark object moving away from the animal (as it did literally during the experiment).

The distribution in the depth of the cortex of most of the units studied, segregated in different columns in the diagram according to their receptive field type, is shown in Fig. 2.8. The only clear laminar segregation observed was for the hypercomplex type "directional to a small object" in the upper half of the cortex.

With respect to other studies on receptive field properties of the rat's striate cortex, the results of Wiensenfeld and Kornel (80) are similar in many aspects to ours. However, their main subdivision of units, as sensitive to stationary or moving stimuli, does not apply to our observations. For example, in Figs. 2.3 and 2.4, it can be seen that the complex cell responded equally well to a moving or a

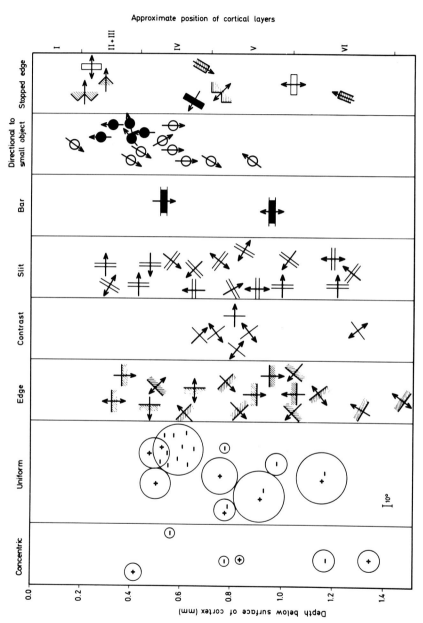

FIG. 2.8. Distribution of the receptive field types of the units studied within the cortical depth. All the receptive fields are drawn to the same scale. The only clear laminar segregation occurred for the hypercomplex "directional to small objects" type in the upper half of the cortex. Arrows indicate directional or nondirectional responses.

stationary edge pattern. Also, they did not describe cells with hypercomplex receptive field properties. The results of Shaw et al. (67) are utterly different from ours, probably because they consider the rat's eye unduly hypermetropic.

The types of receptive fields found in the striate cortex of the rat, as listed above, are as sophisticated as those found in higher mammals, and substantial proportions of them have much more complex properties than the receptive fields of cells in the dorsal lateral geniculate nucleus of the rat (38,43,49). Nonoriented receptive field types, although uncommon but present in the striate cortex of the cat (4,16,21) and monkey (6,23), are frequent in the striate cortex of the mouse (14), hamster (72), rabbit (10,51), tree shrew (28) and opossum (62). Oriented receptive fields, simple, complex and hypercomplex, are present in the striate cortex of all the mammalian species referred to above. It appears, then, that the organization of receptive field types in the striate cortex of mammals has some general features; that is, they are made up of nonoriented fields (concentric and uniform) and oriented fields (simple, complex and hypercomplex). A similar suggestion has been made by Dräger (14) and Rocha-Miranda et al. (62).

4. Striate–Extrastriate Corticocortical Connections in the Rat

In a degeneration study of cortical connections from the rat's striate cortex (42), the results showed that these connections are sent into each one of the extrastriate visual cortical areas that were defined electrophysiologically. Thus, from a lesion in the striate cortex, terminal degeneration fields are distributed into each of the cortical regions that contains the AM, AL, LM, LL, PL and P visual cortical areas, as shown in Fig. 2.9. In Fig. 2.2, these projection fields were indicated by dotted regions in the extrastriate areas. Lesions in different parts of the striate cortex revealed translocations in the projection fields in areas AL and LM that are consistent with the retinotopy of these areas (see Fig. 3 in ref. 42). In Fink-Heimer preparations, the densiest terminal degeneration in all these connections appears in cortical layers II-III, as shown in Fig. 2.10 a, b.

In recent experiments I have traced these connections autoradiographically after injections of ^{3}H-proline in the striate cortex of gray rats. The distribution of labeled fields in the peristriate cortex corroborated the pattern seen with degeneration techniques. That is, after a single injection of ^{3}H-proline in the medial part of the striate cortex (representation of the temporal contralateral vis-

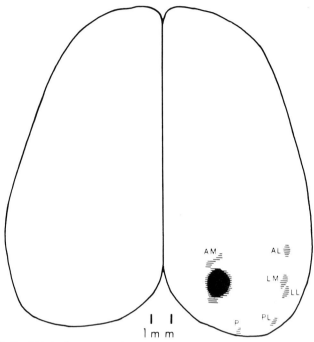

FIG.   2.9. Distribution of cortical projections from the striate cortex
in the peristriate cortical areas of the rat. The black area represents an
electrolytic lesion in the striate cortex. Each line in the hatched regions
represents the terminal degeneration (Fink-Heimer Method) in coronal
sections 90 μm apart.

ual field), labeled fields appear in area P, at the occipital pole, just
below the caudal limit of the striate cortex (Fig. 2.11a). At this caudal level the cortex is cut tangentially, so that the columnar nature
of this projection field is clearly apparent in the micrograph. More
rostrally, the section in Fig. 2.11b shows projection fields to area PL
and to the retrosplenial cortex (RS). The striate cortex projection to
the retrosplenial cortex, readily evident in the autoradiographic
material, is also found in Fink-Heimer material. The labeling at top
left in Fig. 2.11b corresponds to the periphery of the injection site
and that in the white matter to bundles of labeled fibers going
caudally into field P. More rostrally, the section shown in Fig. 2.11c
contains projection fields to areas LM and LL. Denser labeling in
lower and upper parts of the cortex, separated by less labeling in
layer V (see also Fig. 2.11d), reflects the presence of terminals in
these upper and lower cortical layers. Fink-Heimer sections, however, show a clear predominance of terminal degeneration in the
supragranular versus infragranular layers (Fig. 2.10a, b). Even
more rostrally, the section in Fig. 2.11d shows the projection field

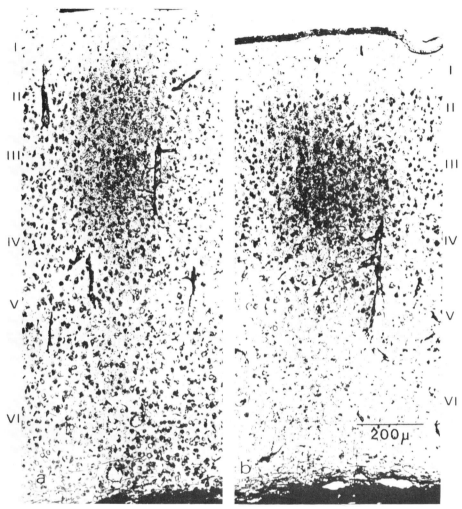

FIG. 2.10. Terminal degeneration in (a) extrastriate area LM and (b) extrastriate area AL after a lesion in the striate cortex of the rat. The densest concentration of degeneration grains is readily evident in the supragranular cortical layers. Roman numerals indicate cortical layers. Fink-Heimer method. (From Montero, Bravo, Fernández, 1973, ref. 42).

into area AL. Note the labeled bundles of fibers in the white matter en route to subcortical projections. The projection field into area AM (not shown) was also present in this autoradiographic material.

In a recent study (56), Olavarria and Montero injected horse-radish peroxidase (HRP) into the striate cortex of gray rats to determine which of the extrastriate areas recipient of striate cortical projections are reciprocally connected with area 17. The results

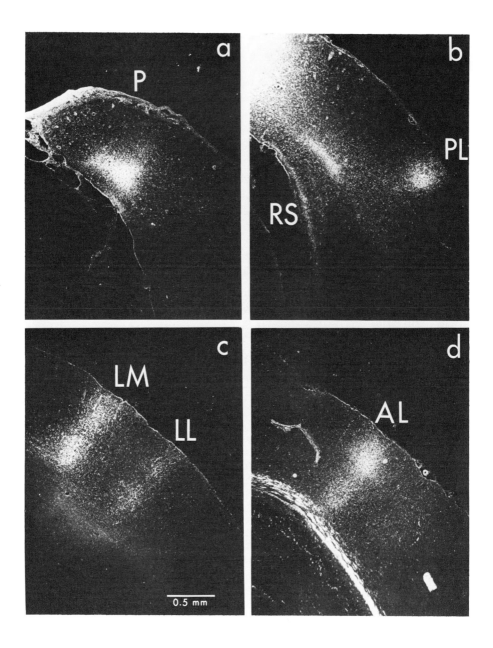

FIG. 2.11. Autoradiographic demonstration of striate cortex projections to the several extrastriate areas in a gray rat. From an injection site
of ^{3}H proline in the medial part of striate cortex (partially seen in b, top
left), there are projection fields to: (a) area P; (b) area PL and to the
retrosplenial cortex, RS; (c) to areas LM and LL; (d) to area AL. The projection to area AM is not shown.

showed that all the recipient extrastriate areas previously defined by degeneration (Fig. 2.9) and autoradiographic methods (Fig. 2.11) showed retrogradely labeled cells. In addition, all these extrastriate regions contained dense accumulations of extraperikaryal grains of reaction products, indicative of anterograde transport of HRP in material processed by the Mesulam protocol (37). The arrangement of these extrastriate fields corresponds closely to the retinotopic subdivision of the peristriate cortex into multiple visual areas (Fig. 2.2), suggesting that each of these areas is reciprocally connected to the striate cortex. In another recent study, Olavarria and Van Sluyters (57a) injected HRP in the superior colliculus of the rat and studied the distribution of labeled cells in the cortex. In cases with superficial collicular injections labeled pyramidal cells were observed in layer V of striate cortex and in several well-defined extrastriate cortical regions. The arrangement of labeled fields in striate and peristriate cortex corresponded closely to the retinotopic subdivision of the rat visual cortex as shown in Fig. 2.2.

In conclusion, the results of anatomical studies with different techniques (degeneration, autoradiography and HRP methods) show that the extrastriate cortex in the rat is subdivided into several different areas that are reciprocally connected to the striate cortex and send connections to the superior colliculus. These extrastriate cortical regions, in turn, correspond to areas that have different retinotopical organizations. All these results, physiological and anatomical, are consistent with the subdivision of the extrastriate cortex in the rat into the several morphofunctional areas, as described in Fig. 2.2.

5. Effects of Postnatal Enucleation of the Eye on the Striate–Extrastriate Connections in the Rat

Having defined the system of striate–extrastriate connections in the rat, we wanted to know the effects of sensorial deafferentation on these connections, as a starting point for further deprivation studies on these pathways. We did this study in collaboration with V. Fernández and M. A. Carrasco (44). Rats had one eye enucleated at 14 days postnatally to avoid the sprouting of the ipsilateral retinogeniculate axons that occurs after early eye enucleation (34). After survival of seven months from the enucleation, electrolytic lesions were placed bilaterally in the monocular segment in the medial striate cortex. The ensuing degeneration in the terminal fields of corticocortical pathways was studied by the Fink-Heimer method 4 days after the cortical lesions. The results showed that, in number and distribution, degenerative projection fields were similar in both hemispheres of the four enucleated animals, as well as in two control animals; i.e., there were projections to cortical areas P, PL, LL, LM, AL and AM. However, on qualitative inspection (Fig. 2.12) and quantitative analysis (number of degeneration grains per unit volume of tissue), there was a striking increase in the density of degeneration grains in the LL field on the undeprived side contralateral to the remaining eye, the density in LL field in the opposite (deprived) side and also with respect to the density in LL field of the control animals. No significant interhemispheric difference in grain density was found in the other cortical fields of degeneration. Density of degeneration in the AM field was difficult to compare between animals owing to its fusion with degeneration surrounding the lesion and was not included in this study.

The quantitative analysis of these results is shown in the histograms in fig. 2.13. The clear and hatched bars represent degeneration grain density for the different fields on the undeprived and deprived sides of enucleated animals, respectively, whereas the dotted bars represent these values from control animals. To compute these values, counts of degeneration grains were taken from 10 samples of 900 μm^3 of tissue at the site of the most dense degeneration in each field. The bars represent the averages of the combined results from the different animals. These results show that density of degeneration grains in the LL field is about three times higher in the undeprived side than in the deprived side or in the control animals. In fact, it almost reached the level of densest degeneration in field LM. We had previously reported (42) that degeneration in field LL was one of the least dense, while that ot field LM was one of the most dense.

These results indicate that enucleation of one eye 14 days after birth induces an abnormal increase in the connectivity of the striate– peristriate pathway to area LL in the hemisphere contralateral to the remining eye. This effect is peculiar for this area. The increase in connectivity is probably related to functions of this particular pathway and area in the intact animal, the nature of which is at present obscure. Speculating on this point, it can be thought *a priori* that the functions that may be increased in an eye-enucleated animal, especially in one with panoramic vision like the rat, are parameters of movements of the remaining eye. Their frequency and extension are probably enhanced in order to compensate for the lack of vision from one eye in the scanning of the visual environment. According to this hypothesis, striate cortex connections to LL would mediate some type of visual function related to eye movements. It would be interesting to know whether a similar increase

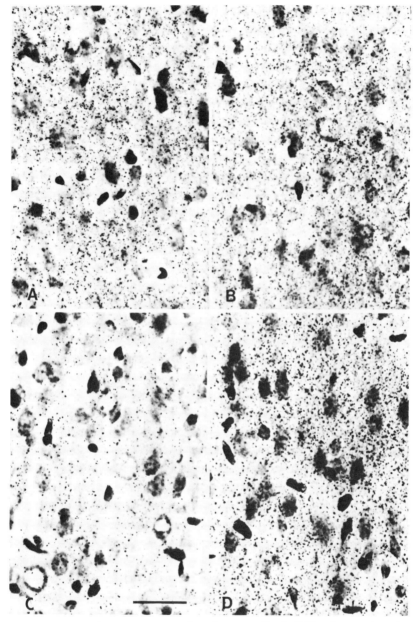

FIG. 2.12. Photomicrographs show on the left the densest regions of terminal degeneration in cortical layers II–III of the LL field on the undeprived side (A) and on the deprived side (C) of an eye enucleated rat; the two photomicrographs on the right correspond to the LM field in the undeprived (B) and deprived sides (D). It can be seen that the degeneration density of the LL field on the undeprived side (A) is considerably greater than on the deprived side (C). It is similar to the density in the LM field (B,D), in which the density is similar on both sides. Bar represents 50 μm.

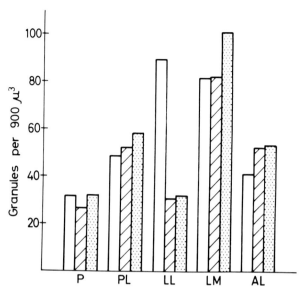

FIG. 2.13. The combined results from the four eye-enucleated animals and the two controls are shown in this histogram. The clear, hatched
and dotted bars are the averages of the values of degenerative density for
the different fields on the undeprived and deprived sides and in the control animals, respectively. These results show that degeneration in the LL
field is about three times denser on the undeprived side than on the deprived side and in the control animals. No significant interhemispheric
density differences exist for the rest of the fields in the enucleated animals
and no differences are seen when comparison is made with control rats.

in the connection from the striate cortex to LL occurs in rats visually deprived by lid suture, since these animals would also be sub-
jected to a similar need to increase parameters of eye movements,
as in animals with enucleation of one eye.

6. Studies on the Visual Cortex of Other Rodents

An electrophysiological mapping of the visual cortex of the degu
(*Octodon degus*) (55), a diurnal rodent with significant ecological
and fossorial differences from the rat, has shown a remarkable similarity in the retinotopic organization of its extrastriate visual areas
to those of the rat. Thus, areas similar to AM, AL, LM and LL were
defined. The regions of areas PL and P were not explored. An additional similarity in the organization of the extrastriate visual cortex
in these two rodents has been reported in an anatomical degenera-

tion study on the degu (30). The distribution patterns of striate–extrastriate projections in this rodent were found to be virtually identical to those in the rat; i.e., the striate cortex sends projections to areas analogous in position to areas P, PL, LL, LM, AL and AM.

As for other rodents, extrastriate visual areas analogous in position to areas AM and LM of the rat have been described in the mouse (14), golden hamster (72) and guinea pig (8). In these studies, however, multiple cortical visual areas in the lateral peristriate cortex were not described, although the reports on the hamster and guinea pig did not explicitly refute their existence. The distribution of visual evoked potentials to gross photic stimulation of the contralateral eye in the mouse suggested the existence of two foci of responses in the occipital cortex, one medial in the striate cortex and the other lateral, possibly in the extrastriate cortex (84). An electrophysiological study of striate and extrastriate cortex in the gray squirrel (19) has described visual areas V II, V III and a temporal area (Tp) in a lateral progression beyond V I in the striate cortex. However, in a recent autoradiographic study in collaboration with K. Cliffer (11), we have found the pattern of striate cortex projections to the peristriate cortex in the gray squirrel to be essentially similar to those of the rat (although most of these projections differ from those of the rat in being segregated in closely related multiple patches). These results in rodents of different suborders suggest a plan of subdivision of extrastriate visual cortical areas, that is general for rodent species of widely different ecology and behavior. We call this a "rodent" type of visual cortex organization.

We would like to discuss in some detail the results of another recent electrophysiological map of the mouse visual cortex by Wagor et al. (78), because of their relevance to the idea of a common "rodent" visual cortex organization. The results reported by these authors are in good agreement in many respects with our map of the extrastriate cortex in the rat (50) in the regions they explored.

(a) They described an area anteromedial to the striate cortex with a preponderance of the temporal visual field representation, quite similar to our findings in area AM in the rat (see Fig. 2.2).

(b) They reported, "in the central part of the lateral extrastriate cortex, there is a second reversal in isoazimuth lines that indicates an additional representation of the nasal lower visual field. Posterior and anterior to this region, such a reversal is not evident; instead, the extreme peripheral fields are represented." This is equivalent to the second reversal of the visual field, from temporal to nasal that occurs at the LM/LL border in the rat's map and, more precisely, it corresponds to the rostral part of area LL, where the lower nasal visual field is represented (see Fig. 2.2). Rostral to this region, the absence of reversal of visual field azimuths in lateral progressions is exactly what we found in area AL of the rat (see Fig. 2.2, area AL).

(c) They found a representation of the upper visual field in the most rostral part of the extrastriate cortex anterolaterally to area 17. This is exactly what we found in the rostral part of area AL in the rat (see Fig. 2.2, area AL). In their paper they did not show explorations of the most caudal part of the occipital cortex, where areas P and PL may be located.

Wagor et al. concluded that the lateral extrastriate cortex in the mouse is composed of areas V2 (V II) and V3 (V III), separated by the 25° elevation line in their map (their assumed "horizontal meridian"). In V2 (V II), the lower and upper visual fields are represented rostrally and caudally, respectively, while in V3 (V III), the upper and lower visual fields are represented rostrally and caudally, respectively. Note that the representation of the upper–lower visual field axis in their area V III of the mouse, is inverted with respect to the representation of this axis in area V III of the cat (13,22,77), in which the lower visual field is rostral. To support the claim that the +25° elevation line in their map corresponds to the horizontal meridian, they did two extrapolations from the results of other authors on the mouse and the rat. They used an approximation of the parameters of the projection of the optic disc in the visual field observed by Dräger (14) (since they did not do this projection) and they assumed the existence of a higher concentration of ganglion cells in the retina of the mouse, which would be in a similar position with respect to the optic disc as the central area in the rat (16).

A study of the striate–extrastriate connections in the mouse would be useful to determine whether their distribution conforms to the subdivisions of extrastriate visual areas as seen in the rat, or whether they conform to the subdivisions of the extrastriate areas as proposed by Wagor et al.

7. Thalamic Afferents to Extrastriate Visual Areas in Rodents

The thalamic afferent connections to different regions of the extrastriate cortex in the rat have been recently studied by several workers using the HRP method. There seems to be agreement that the medial extrastriate cortex, which would contain area AM, receives connections from the rostral part of the lateroposterior nucleus (LPN) (27,59; Olavarria, personal communication). In the mouse and hamster, the medial extrastriate cortex, which would contain area AM, appears to receive connections from the lateral nucleus (LN) (7,15). Olavarria (54) reported that HRP injections in areas AL, LM and LL in the rat, placed after electrophysiological mapping, labeled cells in ventrolateral, central and medial parts, respectively, of LPN (in Fig. 1 of his report, however, it can be seen that the injection in area LL labeled many more cells caudally than rostrally in LPN, with respect to injections in LM and AL). Perry (59) reported that injections placed more laterally in area 18a of rat labeled regions more posterior in LPN. However, in Fig. 7 of his paper, after an injection in the medial part of 18a, located in the region of LM, more cells are labeled in the caudal part of LPN. In this respect, the results of Perry and Olavarria are similar. Hughes (27) and Coleman and Clerici (12) reported that more lateral HRP injections in the occipital cortex of the rat label more caudal parts of LPN. In the gray squirrel, Robson and Hall (61) found that the most caudal part of LPN projects to the temporal cortex, while the rostrolateral and rostromedial parts of LPN project to areas 18 and 19, respectively. In the degu, Kuljis et al. (31) described a similar distribution of connections from LPN to areas 18, 19 and temporal cortex as that reported for the gray squirrel (61). Injections of HRP in the most caudal regions of the extrastriate cortex in the rat (12), possibly containing areas P and PL, label lateral and central regions of LPN.

With respect to superior colliculus (SC) connections to LPN in the rat (59), Perry showed that the caudomedial part of LPN receives bilateral afferents from the SC and that the anterolateral part of LPN receives only ipsilateral afferents from SC. This pattern is similar to that observed in the gray squirrel (61). However, in the rat (59), no part of LPN appears to be devoid of SC afferents (59), in contrast to the squirrel (61), in which a rostromedial part of LPN does not receive SC inputs. Thus, all extrastriate visual areas in the rat would receive and integrate inputs from the retino-geniculo-cortical system (via striate cortex connections) and from extra-geniculate pathways (via retino-colliculo-lateroposterior system). The latter pathway would explain the maintenance of retinotopy in some of the extrastriate areas in the rat after destruction of the striate cortex (57).

The possibility that dLGN projects to the extrastriate cortex in the rat is controversial, since some studies support this projection (12,26), while others do not show or support the existence of this projection (54,59,60). The most convincing evidence, in my opinion, is the autoradiographic tracing of projections from dLGN injected with ^{3}H-amino acid (after a short survival time, which would prevent transneural transport). These results showed that geniculate projections are restricted to the boundaries of the striate cortex in the rat (60).

8. Behavioral Studies on the Extrastriate Cortex of the Rat and Rabbit

In a behavioral experiment done in collaboration with T. Pinto-Hamuy and coworkers (17), thermal lesions, carefully restricted to the lateral group of extrastriate visual areas in rats, induced a slight but significant deficit in visual pattern discrimination and a total failure on a visual conditional discrimination test. These results suggest that the lateral extrastriate areas in the rat have a role in the acquisition of pattern discrimination and are essential in tasks requiring the association of two visual cues.

In a similar experiment on the rabbit, Murphy and Rappaport (52) reported that destruction of the group of lateral extrastriate visual areas in the occipital and temporal cortex induced a failure to acquire a reversal learning set of brightness discrimination, in contrast to normal animals. Normal and operated animals did not differ in their rates of learning and initial brightness discrimination. The results suggest that the extrastriate visual system in the rabbit has the same function in visual learning and learning-set formation as it does in higher mammals.

9. Conclusions from Studies on the Rat

In conclusion, despite its apparent cytoarchitectonic homogeneity in Nissl sections (29), the extrastriate visual cortex of the rat contains several retinotopically defined visual areas, that receive separate direct connections from area 17, and send connections back to the striate cortex. In addition, these areas receive thalamic inputs from LPN. Thus, these are cortical functional units processing striate cortical and extrageniculate thalamic inputs, some of them of collicular origin. Integrity of the lateral group of extrastriate areas in the rat is essential for a successful performance of a visual conditional discrimination test and, in the rabbit, for the acquisition of a learning set. Multiple peristriate areas are also present, in analogous positions with respect to the striate cortex, in other rodents. The functions of each of these visual areas in cortical visual processes remain to be determined. The identification of these areas, on physiological and morphological grounds, represents a first step in this direction. Further electrophysiological, anatomical, behavioral and other types of studies are required to advance in this direction.

10. Cortical Connections from the Striate Cortex in the Rabbit

In a study on the distribution of cortical projections from the striate cortex in the rabbit in collaboration with E. H. Murphy (48), labeled projection fields appeared in several extrastriate cortical regions after single injections of ^{3}H-proline in area 17. An example of results of these experiments is shown in Fig. 2.14. The retinotopy of the cortical injection site, estimated from the map of Thompson et al. (71) and from the position of the projection field on a flattened reconstruction of the superior colliculus according to Hughes (26) (SC in fig. 2.14), was at about $+10°$ of elevation and $30°$ of azimuth. Cortical projection fields are indicated by dotted regions on a lateral view of the brain and on drawings of coronal sections, whose planes are indicated by lines on the brain schema. The injection site is indicated by black in the brain and section drawings.

Projections from the injection site, present only ipsilaterally, were distributed in the following cortical areas, named according to their position relative to the striate cortex as in the rat:

(a) The caudal part of the cortex adjacent ventrally to area 17. This "posterior" field is indicated by P in the brain diagram and in section A of Fig. 2.14.

(b) Two closely related fields were present in the temporal cortex, T2, of Rose (64). These two "posterolateral" fields are indicated by PL' and PL'' in the diagrams of brain and section C of Fig. 2.14.

(c) More rostrally, in the midportion of Rose's area occipitalis, there was a "lateromedial" field indicated by LM in the brain and section diagram of Fig. 2.14. This field is characteristically elongated in the caudorostral direction.

(d) Ventral to LM, but still in area occipitalis, there was a "laterolateral" field labeled LL in the diagrams of brain and sections in Fig. 2.14.

(e) In the rostral part of area occipitalis, there was an "anterolateral" field labeled AL in Fig. 2.14, section G. In the rostral part of Rose's T1, there was an "anterotemporal" field, labeled AT in Fig. 2.14. All these cortical projection fields showed the densest terminal label in the upper half of cortical layers, conveyed via U-fibers that entered the white matter.

(f) In addition, there was an "anteromedial" projection field, via labeled intracortical fibers emanating from the injection site anteromedially (labeled AM in section H in Fig. 2.14) and ending in Rose's area peristriata.

(h) Finally, there were projections to the presubiculum, labeled PS in Fig. 2.14, and projections to the alpha and beta sectors of the retrosplenial cortex.

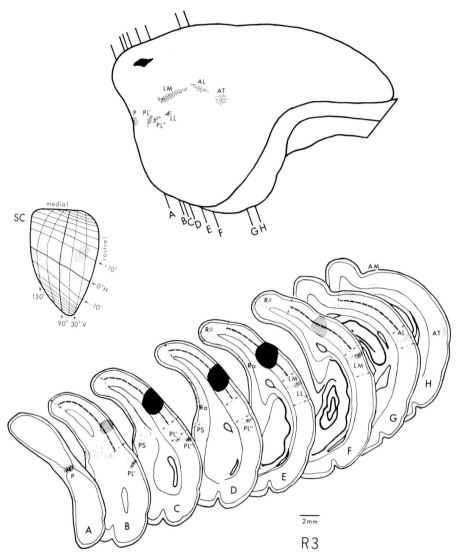

FIG. 2.14. Lateral view of the brain and diagrams of series of sections (A–H) of rabbit brain R3, to show the several labeled fields in the cortex that appear after an injection of ^{3}H-proline (shown by black) in the striate cortex (indicated by interrupted line in the sections). The different projection fields are: P, posterior; PL′ and PL″, posterolateral 1 and 2; LM, lateromedial; LL, laterolateral; AL, anterolateral; AM, anteromedial; AT, anterotemporal; PS, presubiculum; Rα and Rβ, retrosplenial alpha and beta. The occipital area of Rose (64) was delimited by two arrows lateral to area 17 and contains LM, LL and AL fields; PL′ and PL″ are in the temporal cortex (T1 and T2) of Rose. The retinotopy of the injection site, inferred from the position of the labeled field in the superior colliculus (SC) in relationship to the retinotopic map in its flattened surface (Hughes, 26) at about +10° of elevation and 30° of azimuth.

These anatomical observations suggest the existence of several extrastriate visual areas in the rabbit that receive direct connections from the striate cortex. Some of the anatomically defined areas have been explored electrophysiologically. For example, the projectional area AL corresponds in place to the "anterolateral" area described by Woolsey et al. (82). In another report, however, Woolsey et al. (83) described several areas in this region. The projectional area LM in the rabbit appears to correspond to most of area V II of Thompson et al. (71). The projectional area, PL, located in the caudal part of Rose's temporal cortex, into which a single locus of the striate cortex sends multiple divergent connections, corresponds in place to the temporal visual area described by Chow et al. (9). The receptive fields of cells in this area were large in size, sensitive to moving visual stimuli and not clearly arranged retinotopically. The multiple, divergent connections that this area receives from the striate cortex are in keeping with the blurred retinotopy described for this area. This PL or temporal area is probably homologous to area PL in the rat, because of its correspondence in location with respect to the striate cortex and by the large size of their receptive fields (see Fig. 2.2). Electrophysiological description of an area equivalent to the projectional area LL in the rabbit is still lacking. We have distinguished this LL projectional field from PL fields, mainly because of its location in Rose's area occipitalis, instead of in the temporal cortex. In the rat and degu, area LL is clearly defined electrophysiologically by a second reversal of the nasotemporal direction of the visual field lateral to LM. In addition, its receptive fields are not as large as those of area PL in the temporal cortex (see Fig. 2.2). Projectional area P, in the caudal aspect of the occipital cortex, remains to be explored electrophysiologically in the rabbit, as well as in other rodents than the rat. The projectional area AT was inconstantly present in different experiments. We have not seen a similarly located projection in the rat. Physiological exploration of this area in the rabbit is still lacking. The projectional area AM in the rostral part of area peristriata of Rose, is similar in location to area AM of rat and other rodents. Physiological identification of area AM in the rabbit is still lacking.

In a degeneration study of corticocortical connections of rabbit's visual cortex, using Fink-Heimer and other variations of the Nauta method, Towns et al. (74) reported a much more restricted pattern of cortical connections from the striate cortex than those reported in our study. These authors reported that regions of the lateral striate cortex send connections to an adjcent portion of the medial occipital cortex and to a restricted region of the lateral occipital cortex. The difference between their results and ours can be only partially attributed to the use of a different technique, since, for instance, Mathers et al. (36) and Bousfield (5), using Fink-

Heimer technique or its Wiitanen modification, have corroborated the striate cortex connections that we have described to the caudal part of Rose's temporal cortex. However, for some unknown reason, the Fink-Heimer technique appears to be unreliable for tracing degeneration in the rabbit corticocortical system [personal experience; Mathers et al.(36)]. Here it is worthwhile to mention that H. Holländer has observed a pattern of striate–extrastriate connections in the rabbit, in autoradiographic material, that is similar to our results (personal communication).

In conclusion, the similarities in the pattern of distribution of the cortical projections from the striate cortex in the rat and rabbit are numerous and striking and are unlikely to arise from simple coincidence. They probably reflect a common or basic plan of organization of these connections and of the visual cortex in rodents and lagomorphs. Still, we label this a "rodent" type of organization of striate cortical projections, in consideration of the greater abundance and variety of rodent species in comparison to lagomorphs.

11. Cortical Connections from the Striate Cortex in the Cat

In a recent study (41), I have traced the topographic distribution of cortical connections from the striate cortex in the cat. Injections of radioactive amino acids were placed in regions of the striate cortex representing central, intermediate and peripheral parts of the horizontal meridian and also in regions of lower and upper visual field representations near the vertical meridian. The cortical projections from these injection sites were studied in autoradiographic preparations. A brief account of these results is given in this review.

Several retinotopic arrangements in the distribution of these cortical connections were differentiated which, it is argued, reflect the retinotopy of the extrastriate cortical recipient areas. The assumption is that these connections are established with retinotopically equivalent cortical regions, in the way it has been shown for striate cortex connections with subcortical centers, such as the lateral geniculate nucleus and superior colliculus (18,46) and with some cortical areas (40).

Figure 2.15 illustrates the results of an experiment in which the rostral part of the striate cortex was injected with ³H-proline. The injected site represents a region in the lower visual field at about $-20°$ of elevation and about 5° of azimuth, estimated from the map of Tusa et al. (76) and from the relative position of the projection field in the superior colliculus with respect to its retinotopy

(70). A given cortical field is indicated by identical numbers in the dorsal view of the brain and in the diagrams of the coronal sections.
The following cortical projections fields were found:

(a) Field 1, in the rostral part of area 18, is on the crown of the lateral gyrus.
(b) Multiple labeled patches, fields 2–5, in area 19 on the medial wall of lateral sulcus.
(c) Fields 6 and 7 are in the cortex of the medial wall and fundus of the middle suprasylvian sulcus.

The results of this experiment show that cortical projections from a region representing lower visual field near the vertical meridian are distributed in rostral parts of areas 18, 19 and suprasylvian belt (Ss) cortex (65), reflecting the retinotopy of visual areas that will be referred to here as V II, V III and LS (for lateral suprasylvian area), respectively.

In another experiment, in which the injection site in area 17 was in a mapped area representing −5° of elevation and 10° of azimuth (not illustrated, but see case A201 in ref. 44), fields in areas 18 (V II), 19 (V III) and Ss (LS) were all displaced caudally in these areas with respect to those fields in case B41.

Projections from a striate cortex region representing a part of the upper visual field, at about +10° of elevation and 1–2° from the vertical meridian are shown in the experiment illustrated in Fig. 2.16. All cortical projection fields are located in more caudal parts of the cortex than fields in case B41 (and case A201). Field 1 is in area 18 and corresponds to V II. Field 2 is in the lateral part of area 19 and corresponds to V III. Fields 3–8 are in the suprasylvian belt cortex (Ss). Fields 5–8 are multiple projections located in the most caudal part of both banks of the middle suprasylvian sulcus and are interpreted to pertain to area LS. Fields 3 and 4, located at the junction of regions Ssp and SSv of Sanides and Hoffmann (65), are interpreted to pertain to a projectional area that exists in the posterior suprasylvian sulcus, area Ps (see below).

The cortical distribution of striate cortex projections, as seen in these experiments, indicates that lower and upper visual fields are represented rostrally and caudally, respectively, in areas V II, V III and LS.

Cortical projections from a region in area 17 representing 15° of azimuth at the zero horizontal meridian, as determined by microelectrode mapping, are illustrated in Fig. 2.17. Field 1, deep on the medial bank of the lateral sulcus is in area 18 and corresponds to V II. Fields 2–4 are a fragmented projection in area 19 and correspond to V III. The multiple projections 10–13 on the medial bank of the middle suprasylvian sulcus correspond to area LS.

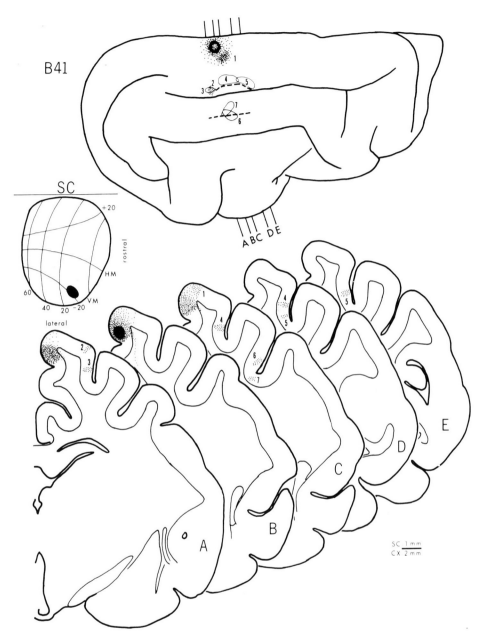

Fig. 2.15. Cortical and collicular projection fields from a single injection site in a region of the striate cortex of cat with an estimated receptive field at 5° of azimuth and −15° to −20° of elevation (cat B41). The region of heavily labeled cells at the injection site is shown in black in coronal section B and its orthogonal projection on the dorsal view of the brain is shown by the thick-lined circle. In the brain diagram, the labeled cortical projection fields that are seen on the cortical surface from this dorsal view (e.g., fields 1,2), are indicated by dotted regions, whereas those in depth are indicated orthogonally by encircled areas.

In addition to the above projections, which belong to areas already defined in the previous experiments (V II, V III, LS), there are other projections that do not conform to the emerging retinotopic trends of these areas. These are fields 8 and 9, on the posterior bank of the posterior suprasylvian sulcus (area PS), fields 6 and 7, deep on both banks of the posterolateral sulcus, and a small field 5 on the lateral gyrus. Note that the last two groups of fields (6–7 and 5) are additional representations of the horizontal meridian at 15° azimuth, that coexist with the projections already described for V II and V III above.

In case C15, illustrated in Fig. 2.18, the receptive field at the injection site, deep on the dorsal bank of the splenial sulcus, was at 45° of azimuth and in the horizontal meridian (HM). That is, the relative displacement of the projection fields in this animal, with respct to those of the previous case B224, would represent a translocation of 30° along the HM towards the perphery. Field 4, at the 18/19 boundary in the fundus of the lateral sulcus is interpreted to represent the confluence of projection fields of the HM of areas V II and V III, at 45° of azimuth. Note that projections of the HM at 15° of azimuth in V II and V III, and fields 1 and 4 in case B224 (Fig. 2.17), are widely separated on both banks of the lateral sulcus at a similar coronal plane. The group of highly segregated fields numbered 8 on the dorsal lip of suprasylvian sulcus, which form multiple parallel bands in consecutive sections, represents the periphery (45°) of the HM in area LS. Note that fields 10–13, in case B224, are situated more deeply on the medial bank of the sulcus, at about the same coronal plane as fields 8 in case C15, and represent intermediate regions (15°) of the HM of area LS. By the same token, fields 5–6 in case C15 pertain to the HM of area PS, but represent its more peripheral part than fields 8–9 in case B224. Also, fields 1–3, laying in an extensive region in the ventral part of the postlateral gyrus, pertain to the same HM representation, but to a more peripheral part than fields 6–7 deep in the postlateral sulcus in case B224. Since this HM representation coexists with other, separate HM fields in areas V II and V III, it will be referred to here as the posterior field of

---

Fig. 2.15 (*continued*): Each number 1–7) indicates the same projection field on both the dorsal view of the brain and in the coronal sections. The interrupted line in the brain diagram represents the fundus of the middle suprasylvian sulcus and lines A–E at bottom and top of brain diagram indicate planes of the coronal sections shown below. The projection field in the superior colliculus (SC) is shown in black. Scale represents 2 mm for both the brain diagram and sections (CX) and 1 mm for SC. These conventions are maintained in the following figures. (from Montero, 1981, ref. 41).

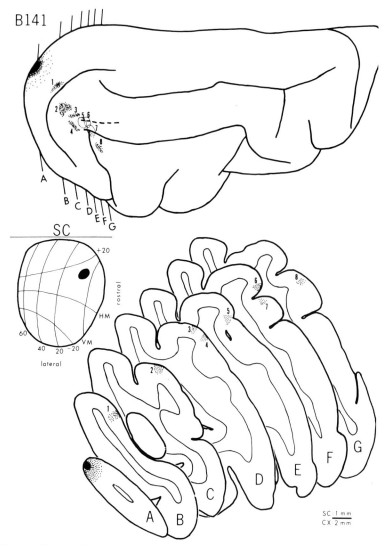

FIG. 2.16. Cortical and collicular fields in cat B141. The receptive field at the injection site is estimated to be at about +10° of elevation and 1–2° from the vertical meridian. Format and abbreviations same as in Fig. 2.15 (from Montero, 1981, ref. 41).

the HM (HMp in Fig. 2.21). Finally, field 7 in case C15 pertains to the same HM representation as field 5 in case B224 and will be referred to here as the anterior field of the HM (HMa in Fig. 2.21) for the same reason given above for HMp.

Cortical projections from striate cortex containing representations from central and paracentral visual fields were analyzed in the

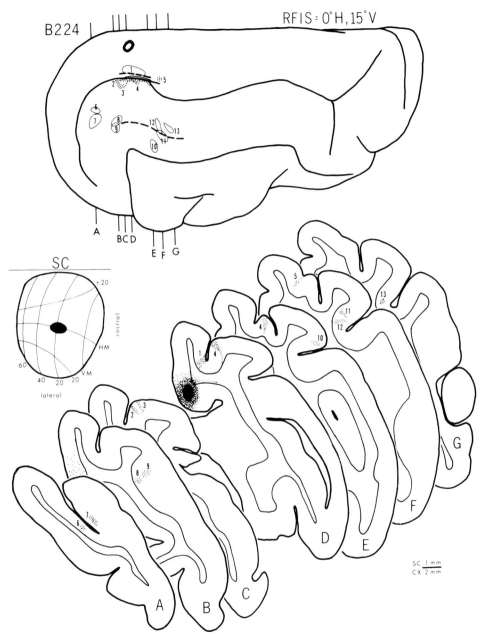

FIG. 2.17. Cortical and collicular projection fields for cat B224. The mapped receptive field at the injection site (RFIS) was at the horizontal meridian and 15° of azimuth. Format and abbreviations same as in Fig. 2.15 (from Montero, 1981, ref. 41).

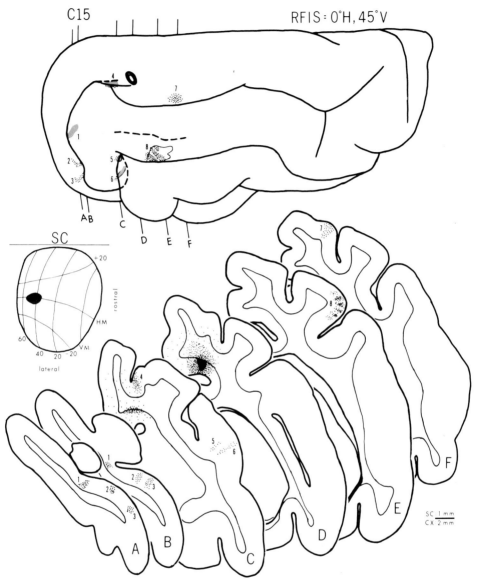

FIG.   2.18. Cortical and collicular projection fields for cat C15. The
mapped receptive field at the injection site (RFIS) was at the horizontal
meridian and 45° of azimuth. Format and abbreviations same as in Fig.
2.15, with the exception that some of the fields that are seen in depth in
the brain diagram are represented by hatched areas (from Montero, 1981,
ref. 41).

results of an experiment, in which the striate cortex was injected
with two different isotope amino acids in two places (case C88, Fig.
2.19). Locations of the injection sites in the visual field representation in the striate cortex were estimated from the map of Tusa et al.

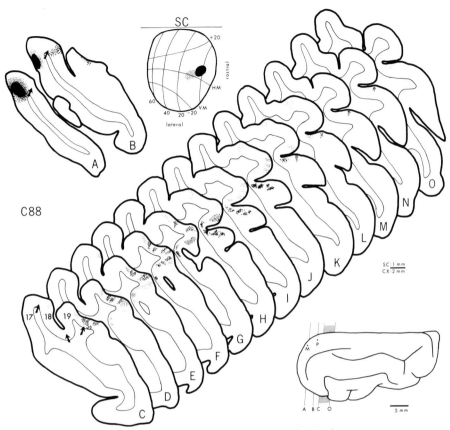

FIG. 2.19. Cortical and collicular projection fields from two separate injection sites (M and P) in cat C88. In the brain diagram, the dots M and P indicate the points of entry of ^{35}S-methionine and ^{3}H-proline injections, respectively. The region of heavily labeled cells at the injection site of ^{35}S-methionine is shown in black in section A, and for ^{3}H-proline in section B (at upper left). The cortical projection fields containing ^{35}S or ^{3}H are indicated by coarse and fine dots, respectively, in coronal sections A–O. The collicular projection field containing ^{35}S is represented by the black area in SC and that containing ^{3}H by the dotted area in SC. The border between areas 17 and 18 is indicated by arrows in sections A, B, and C. In section C, other arrows indicate borders between areas 18 and 19, and between areas 19 and Ss. Lines A–O in the brain diagram indicate planes of the coronal sections above (from Montero, 1981, ref. 41).

(76). The injection site of ^{35}S-methionine was located at about 1° from the vertical meridian and +1° to +2° of elevation. The receptive field at the ^{3}H-proline injection site was estimated at about 3° to 4° of azimuth on the horizontal meridian.

The central–paracentral relationship of the ^{35}S-methionine and ^{3}H-proline injection sites is clearly expressed in the relation of the ^{35}S (black area) and ^{3}H (dotted area) projection fields in the SC diagram in Fig. 2.19. The ^{35}S field is adjacent rostrally, and slightly medially, to the ^{3}H field. Conversely, this topographical relation of the collicular fields is direct evidence of the strict retinotopic organization of the striate corticotectal connections and of their equivalence to collicular retinotopy.

It should be emphasized, before describing the cortical projections of case C88, that the advantageous feature of double label autoradiography (45) is that it is possible to establish topographical relationships of the projections from two injection sites in the same tissue, when there is spatial separation of the two types of labeled terminals. This has been shown before for striate cortex projections to the thalamic reticular nucleus (47) and to the striate cortical recipient area in the superior temporal sulcus cortex of the rhesus monkey (40). Thus, the spatial relations of ^{35}S- and ^{3}H-label occurring in different cortical areas in cat C88 define a central–paracentral topographical gradient for these connections.

In area 18, projection fields labeled with ^{35}S (section B) and ^{3}H (sections C–E) are adjacent in a caudomedial to rostrolateral direction, at the junction of the lateral and posterolateral gyri.

In area 19, the fields labeled with ^{35}S and ^{3}H follow a caudolateral–rostromedial gradient (sections B–L) that begins in the lateral part of area 19 at the junction of the suprasylvian and posterosuprasylvian gyrus.

Beyond area 19, there is a complex relationship betweeen ^{35}S and ^{3}H projections in the cortex, corresponding to the suprasylvian sulcus belt (Ss.). Caudally (in sections C–F), there is a mirror image reversal of ^{35}S and ^{3}H fields in Ss with respect to those seen in area 19, despite the fact that both types of projections are segregated into multiple patches or bands. That is, ^{3}H fields are lateral to ^{35}S fields and they occupy the upper part of the caudal bank of the posterior suprasylvian sulcus. (In this context, "lateral" and "medial" are used to refer to regions farther from and nearer to area 17, respectively, along the contour of the cortex in coronal sections). Rostrally, in sections H–K, ^{35}S projections move progressively laterally into the lateral bank of the middle suprasylvian sulcus. Further rostrally, in sections L–O, ^{3}H fields move onto the lateral bank of the sulcus.

The distribution of the ^{35}S fields on the lateral bank and of ^{3}H fields on the medial bank of the suprasylvian sulcus in section J are shown in the dark-field microphotograph in Fig. 2.20. Note the distinctive patchiness of both projections and the expected denser labeling of ^{35}S fields in the lateral bank.

In conclusion, the results of the double label experiment in cat C88 reveal several "central–paracentral" gradients in several different extrastriate areas. One projection is within area 18 and repre-

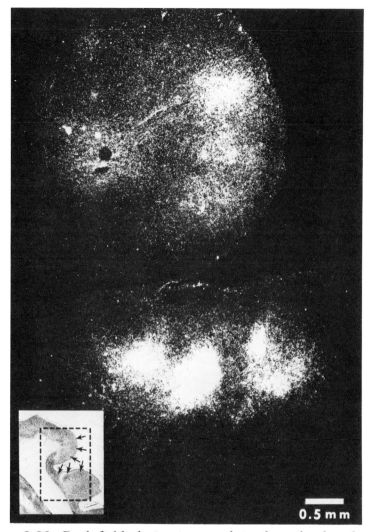

FIG. 2.20. Dark-field photomicrograph to show the distribution of [3]H- and ^{35}S-labeled fields in the medial and lateral banks, respectively, of the middle suprasylvian sulcus in cat C88 (also seen in section J of Fig. 2.19). The arrows in the small bright-field photomicrograph indicate position of labeled fields in both banks of the sulcus. The dashed lined rectangle in this small insert indicates the area shown in the dark-field photograph. Note the segregation of both ^{3}H and ^{35}S patches in the lateral bank. Double emulsion autoradiographic section counterstained with cresyl violet (from Montero, 1981, ref. 41).

sents the central–paracentral region of the retinotopic organization defined above as V II. This gradient is consistent with the mediolateral progression in the lateral gyrus of horizontal meridian fields (from central to periphery) in cats B224 and C15. Also, it is consistent with the rostrocaudal progression in the lateral gyrus of projection fields representing -20° and -5° (cats B441 and A201), respectively.

The second gradient is distributed in area 19 and represents the central–paracentral region of the retinotopic organization defined above as V III. This projection is consistent with the lateromedial gradient of representations towards the periphery in fields in the suprasylvian gyrus (cases B224, C15) and with the rostrocaudal trend in V III of fields moving up in the lower visual field (cases B41, A201).

A third gradient, distributed mediolaterally (or dorsoventrally) in the upper part of the posterior suprasylvian gyrus, represents the central–paracentral region of the retinotopic organization defined above as PS. It is consistent with the dorsoventral trend towards the periphery of fields along the posterior suprasylvian gyrus in cases B224 and C15.

A fourth gradient, distributed lateromedially (or ventrodorsally) along both banks of the middle suprasylvian sulcus, represents the central–paracentral region of the retinotopic organization defined above as LS. It is c onsistent with the ventrodorsal distribution towards the periphery of fields on the medial bank of the suprasylvian sulcus in cats B224 and C15. It is also consistent with the rostrocaudal progression of fields in the suprasylvian sulcus, toward the upper visual field, seen in cases B41 and A201.

The results presented above show that connections from different parts of the striate cortex are distributed topographically to several other cortical areas in the cat. The pattern of distribution of these corticocortical connections reflects the retinotopy of each of these target areas, in the sense that different parts of the striate cortex connect with retinotopically equivalent regions of cortical and subcortical centers, and was used to map anatomically the general retinotopic trends of these areas.

The general retinotopy of the different extrastriate areas, as defined in this study on the basis of the topography of their striate cortical afferent connections, is shown in a summary diagram in Fig. 2.21. In this diagram, for the different areas, the vertical meridian is indicated by continuous lines and the horizontal meridian by arrowlike interrupted lines, in which point to tail directions indicate central to peripheral gradients. Upper and lower visual fields are indicated by the letters U and L, respectively. V I represents the well established retinotopy of the striate cortex (76,81), in which the injections were made. In V II, the horizontal meridian runs laterally on the lateral gyrus, from central to peripheral, dividing regions of lower (rostral) and upper (caudal) visual fields. In V III, the

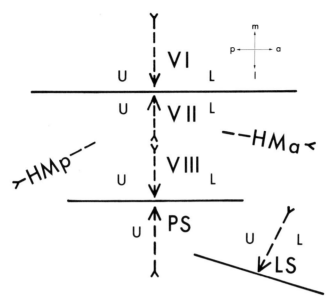

FIG. 2.21 Summary diagram of the general retinotopic distribution of the corticocortical connections from the striate cortex (and of the inferred retinotopy of the extrastriate cortical recipient areas) in the cat, as defined in this study. The diagram is a highly schematized representation of the unfolded cortex, in which the anteroposterior and mediolateral relations of the different cortical areas in the brain are grossly maintained. The mediolateral and anteroposterior axes are indicated by the crossed arrows at top right. V I represents the retinotopy of the striate cortex, in which the injections were made. V II, V III, PS LS, HMp and HMa represent the different retinotopic distributions of connections in the extrastriate areas. In these, the vertical meridian is represented by continuous lines, and the horizontal meridian by arrowlike interrupted lines, in which the point-to-tail direction indicates a center to periphery gradient along this meridian. U and L indicate upper and lower visual field representations. For further explanations, see text (from Montero, 1981, ref. 41).

horizontal meridian, which connects at the periphery (45°) with the horizontal meridian of V II, runs laterally in the suprasylvian gyrus from peripheral to central, dividing regions of lower visual field (rostrally) from regions of the upper visual field (caudally). There are two additional representations of the periphery of the horizontal meridian in the lateral border of area 18, which are different and separate from the horizontal meridian of V II and V III. One is located posteriorly in the ventral part of the posterolateral gyrus and is labeled HMp in the diagram; the other is located anteriorly in the lateral gyrus and is labeled HMa. V III connects laterally, through the area of central vision, with area PS. In area PS, the horizontal

meridian runs ventrally on the caudal bank of the posterior suprasylvian sulcus from central to peripheral; in this area there are projections from the injected regions of upper visual field, but there are
no projections from the regions of lower visual fields injected
(-5°,-20°). In area LS, the horizontal meridian runs ventral to dorsal
along the banks of the middle suprasylvian sulcus from central to
peripheral, dividing regions of lower visual field, rostrally, from regions of upper visual field, caudally.

     For an extensive discussion of comparison of the above de-
scribed general retinotopies of the extrastriate cortical areas in the
cat, with several electrophysiological maps of these areas (2, 3, 13,
22, 24, 33, 35, 58, 75, 77, 79, 81), the reader is referred to the original report of these data (41).

     It is pertinent to comment here, however, that the type of ana-
tomical map reported above offers both advantages and disadvan-
tages with respect to electrophysiological maps. One disadvantage
is that, in a given animal, one can obtain data on the projection of
only one or two retinotopic points of the striate cortex and, it is
argued, of the visual field representation in these areas. Thus, it
was necessary to inject different regions of the striate cortex in several animals to obtain information about the layout of retinotopic
trends at quadrantic levels in these areas. The amount of detail of
retinotopy of a given area that can be obtained in an electrophysiological experiment is limited only by the endurance of the animal
or, more often, of the experimenter. However, the great merit of this
type of anatomical map is that it is possible to obtain in a single animal information about *all* cortical regions in different areas that receive connections from a given point of area 17 and that have, consequently, a representation of that particular retinotopic sector.
Furthermore, this method can give information about geometrical
transformations in the cortical connections from area 17, such as
the phenomena of divergency and convergency seen in striate cortex connections to area LS in the cat and similarly seen in areas
STS of macaque and MT of owl monkey (40). As in the case of the periodic   segregation   of   ocular   dominance   columns   in   the
geniculocortical projection in macaque (25) and cat (68), the
quasiperiodic segregation of striate cortex connections to LS, MT
and STS may be of functional significance in the transfer of information from the striate cortex to these areas. We have advanced the
hypothesis (40,41) that they represent segregation of different
directionality preferences, which are all mixed up in a given striate
cortex locus (21), into "directional columns" in these areas. One
unifying functional property of areas LS, STS and MT is that the receptive fields of their neurons are highly sensitive to motion and
directionality of visual stimuli (24, 53, 68, 85, 86). There is evi-

dence suggesting the existence of directional columns in area STS of macaque (85) and owl monkey (86) and of directional and orientation columns in area LS of the cat (24). Further, there is evidence that directional properties of neurons of area LS of the cat are dependent on striate cortex inputs (69).

The projectional pattern of striate–extrastriate connections in the cat is different in so many respects from that seen in the rat and rabbit, that it is difficult, if not impossible, to suggest analogous systems of connections in these animals, with the exception perhaps of striate cortex connections to areas V II and V III in the cat and to areas LM and LL in the rat and rabbit, for reasons of similarity in retinotopy and position with respect to the striate cortex. Some of the many differences, however, are as follows:

> (a) In the cat there are cortical visual areas that have been mapped electrophysiologically, as, for example, areas AMLS and ALLS described by Palmer et al. (58) located in the rostral part of the middle suprasylvian sulcus, that do not appear to receive direct connections from area 17. In contrast, all the electrophysiological subdivisions in the total extension of the visually responsive cortex in the rat (Figs. 2.1 and 2.2) receive separate, direct connections from area 17 (Fig. 2.9).

> (b) In the cat material, we have never seen striate cortex projections to the retrosplenial cortex and presubiculum that are present in the rat and rabbit (and also in the gray squirrel, ref. 11).

In conclusion, in the cat there are no cortical projectional areas that, for reasons of their position relative to the striate cortex or of their retinotopy, we might compare to areas P, PL, AL and AM of rat and rabbit. The differences between the pattern of cortical connections from the striate cortex in the cat and those of rat and rabbit may be the result of variations appearing in parallel evolution of their ancestors. The organization of these cortical connections in the cat may reflect a general carnivore's pattern of these connections.

Acknowledgments

Studies considered in this review were supported by NIH grants EY-02877, HD-03352, NS-03640, NS-03641, NS-06662 and NS-12470. The technical assistance of Roberto Quiroz and Rene Roi, in the early experiments in my laboratory in Chile, is belatedly acknowledged. For the experiments done in Wisconsin, I acknowledge the technical assistance of Elaine Langer, Shirley Hunsaker and James Van Gemert. For the transcription of this manuscript, I thank Caroline Clark and my wife Olga.

References

1. ADAMS, A. D., AND FORRESTER, J. M. The projection of the rat's visual field on the cerebral cortex. *Quart. J. Exp. Physiol.* 53: 327–336, 1968.
2. ALBUS, K., AND BECKMAN, R. Second and third visual areas of the cat: interindividual variability in retinotopic arrangement and cortical location. *J. Physiol., London,* 299: 247–276, 1980.
3. BILGE, M., BINGLE, A., SENEVIRATNE, K. N., AND WHITTERIDGE, D. A map of the visual cortex in the cat. *J. Physiol., London,* 191: 116–118, 1967.
4. BISHOP, P. O., AND HENRY, G. H. Striate neurons: receptive field concepts. *Invest. Ophthal.,* 11: 346–354, 1972.
5. BOUSFIELD, J. D. Some properties of extrastriate visual units in the cortex of the rabbit. *Brain Res.,* 149: 365–378, 1978.
6. BULLIER, J., AND HENRY, G. H. Ordinal position and afferent input of neurons in monkey striate cortex. *J. Comp. Neurol.,* 193: 913–935, 1980.
7. CAVINESS, V. S., AND FROST, D. O. Thalamic projections to the neocortex in the mouse. II. Spatial order. *J. Comp. Neurol.,* (in press).
8. CHOUDHURY, B. P. Retinotopic organization of the guinea pig's visual cortex. *Brain Res.,* 144: 19–29, 1978.
9. CHOW, K. L., DOUVILLE, A., MASCETTI, G., AND GROBSTEIN, P. Receptive field characteristics of neurons in a visual area of the rabbit temporal cortex. *J. Comp. Neurol.,* 171: 135–146, 1977.
10. CHOW, K. L., MASLAND, R. H., AND STEWART, D. L. Receptive field characteristics of striate cortical neurons in the rabbit. *Brain Res.,* 33: 337–352, 1971.
11. CLIFFER, K., AND MONTERO, V. M. Cortico-cortical connections from the striate cortex in the gray squirrel (*Sciurus carolinensis*) (in preparation)
12. COLEMAN, J., AND CLERICI, W. J. Extrastriate projections from thalamus to posterior occipito-temporal cortex in rat. *Brain Res.,* 194: 205–209, 1980.
13. DONALDSON, I. M. L., AND WHITTERIDGE, D. The nature of the boundary between cortical visual areas II and III in the cat. *Proc. Roy. Soc. London B.,* 199: 445–462, 1977.
14. DRÄGER, U. G. Receptive fields of single cells and topography in mouse visual cortex. *J. Comp. Neurol.,* 160: 269–290. 1975.
15. DURSTELER, M. R., BLAKEMORE, C., AND GAREY, L. J. Projections to the visual cortex in the golden hamster. *J. Comp. Neurol.,* 183: 185–204, 1979.

16. FUKUDA, Y. A three-group classification of rat retinal ganglion cells: histological and physiological studies. *Brain Res.*, 119: 327–344, 1977.

17. GALLARDO, L., MOTTLES, M., VERA, L., CARRASCO, M. A., TORREALBA, F., MONTERO, V. M., AND PINTO-HAMUY, T. Failure by rats to learn a visual conditional discrimination after lateral peristriate cortical lesions. *Physiol. Psychol.*, 7: 173–177, 1979.

18. GAREY, L. J., JONES, E. G., AND POWELL, T. P. S. Interrelationships of striate and extrastriate cortex with the primary relay sites of the visual pathway. *J. Neurol. Neurosurg. Psychiat.*, 31: 135–157, 1968.

19. HALL, W. C., KAAS, J. H., KILLACKEY, H., AND DIAMOND, I. T. Cortical visual areas in the grey squirrel (*Sciurus carolinensis*): A correlation between cortical evoked potential maps and architectonic subdivisions. *J. Neurophysiol.*, 34: 437–452, 1971.

20. HENRY, G. H., HARVEY, A. R., AND LUND, J. S. The afferent connections and laminar distribution of cells in the cat striate cortex. *J. Comp. Neurol.*, 187: 725–744, 1979.

21. HUBEL, D. H., AND WIESEL, T. N. Receptive fields, binocular interaction and functional architectures in the cat's visual cortex. *J. Physiol., London,* 160: 106–154, 1962.

22. HUBEL, D. H., AND WIESEL, T. N. Receptive fields and functional architecture in two non-striate visual areas (18 and 19) of the cat. *J. Neurophysiol.*, 28: 229–289, 1965.

23. HUBEL, D. H., AND WIESEL, T. N. Receptive fields and functional architecture of monkey striate cortex. *J. Physiol., London,* 195: 215–243, 1968.

24. HUBEL, D. H., AND WIESEL, T. N. Visual area of the lateral suprasylvian gyrus (Clare-Bishop area) of the cat. *J. Physiol., London,* 202: 251–260, 1969.

25. HUBEL, D. H., AND WIESEL, T. N. Laminar and columnar distribution of geniculocortical fibers in the macaque monkey. *J. Comp. Neurol.*, 146: 421–450, 1972.

26. HUGHES, A. Topographical relationships between the anatomy and physiology of the rabbit's visual system. *Docum. Ophthal.*, (Den Haag), 30: 33–159, 1971.

27. HUGHES, H. C. Anatomical and neurobehavioral investigations concerning the thalamo cortical organization of the rat's visual system. *J. Comp. Neurol.*, 175: 311–336, 1977.

28. KAUFMAN, P., WALLINGFORD, E., OSTDAHL, R., AND SOMJEN, G. Receptive field properties of visual cortical neuron in the tree shrew (*Tupaia glis*). *Soc. Neurosc. Third Ann. Meet. Abstracts*, p. 178, 1973.

29. KRIEG, W. J. S. Connections of the cerebral cortex. I. The albino rat. A. Topography of the cortical areas. B. Structure of the cortical areas. *J. Comp. Neurol.*, 84: 221–275, 277–323, 1946.

30. KULJIS, R. O., FERNÁNDEZ, V., AND BRAVO, H. The visual system of the *Octogon degus*: normal connection pattern and its reorganiza-

tion following monocular postnatal deafferentation. *Neurcirugia, (Chile),* (in press).

31. KULJIS, R., KAUFMANN, W., FERNANDEZ, V., BRAVO, H., AND FUENTIS, I. Estudio experimental de las aferencias visualis corticalas extrageniculadas en el *Octodon degus. Rev. Med., Chile,* 107: 197–202, 1979.

32. LE MESSURIER, D. H. Auditory and visual areas of the cerebral cortex of the rat. *Fed. Proc.,* 7: 70–71, 1948.

33. LEVENTHAL, A. G., HALE, P. T., AND DREHER, D. The representation of the visual field in areas 18 and 19 of the cat. *Proceed. Austral. Physiol. Pharmacol. Soc.,* p. 59P, 1978.

34. LUND, R. D., CUNNINGHAM, T. J., AND LUND, J. S. Modified optic projections after unilateral eye removal in young rats. *Brain Behav. Biol.,* 8: 51–72, 1973.

35. MARKUSTKA, J. Visual properties of neurons in the posterior suprasylvian gyrus of the cat. *Exp. Neurol.,* 59: 146–161, 1978.

36. MATHERS, L. H., DOUVILLE, A., AND CHOW, K. L. Anatomical studies of a temporal visual area in the rabbit. *J. Comp. Neurol.,* 171: 147–156, 1977.

37. MESULAM, M. M. Tetramethyl benzidine for horseradish peroxidase neurohistochemistry: a non-carcinogenic blue reaction product with superior sensitivity for visualizing neural afferents and efferents. *J. Histochem. Cytochem.,* 26: 100–117, 1978.

38. MONTERO, V. M. Receptive fields of cells in the dorsal lateral geniculate nucleus of the rat. Abstract in *Science* 158: 952, 1967.

39. MONTERO, V. M. Evoked responses in the rat's visual cortex to contralateral, ipsilateral and restricted photic stimulation. *Brain Res.,* 53: 192–196, 1973.

40. MONTERO, V. M. Patterns of connections from the striate cortex to cortical visual areas in superior temporal sulcus of macaque and middle temporal gyrus of owl monkey. *J. Comp. Neurol.,* 189: 45–59, 1980.

41. MONTERO, V. M. Topography of the cortico-cortical connections from the striate cortex in the cat. *Brain Behav. and Evol.,* (in press), 1981.

42. MONTERO, V. M., BRAVO, H., AND FERNÁNDEZ, V. Striate– peristriate cortico-cortical connections in the albino and gray rat. *Brain Res.,* 53: 202–207, 1973.

43. MONTERO, V. M., AND BRUGGE, J. F. Direction of movement as the significant stimulus parameter for some lateral geniculate cells in the rat. *Vision Res.,* 9: 71–88, 1969.

44. MONTERO, V. M., CARRASCO, M. A., AND FERNÁNDEZ, V. Effect of postnatal enucleation of the eye on cortico-cortical connections of the rat's striate cortex. *Neurosc. Abstr.,* 1: 75, 1975.

45. MONTERO, V. M., LANGER, E., AND GUILLERY, R. W. Double emulsion autoradiography for differentiating terminals labeled with 3H and 35S. *Soc. for Neurosc. Short course Syllabus: Neuroanatomical Techniques,* p. 47–48, 1978.

46. MONTERO, V. M., AND GUILLERY, R. W. Degeneration in the dorsal lateral geniculate nucleus of the rat following interruption of the retinal or cortical connections. *J. Comp. Neurol.,* 134: 211–242, 1968.

47. MONTERO, V. M., GUILLERY, R. W., AND WOOLSEY, C. N. Retinotopic organization within the thalamic reticular nucleus demonstrated by a double label autoradiographic technique. *Brain Res.,* 138: 407–421, 1977.

48. MONTERO, V. M., AND MURPHY, E. H. Cortico-cortical connections from the striate cortex in the rabbit. *Anat. Rec.* 183: 483, 1976.

49. MONTERO, V. M., AND ROBLES, L. Saccadic modulation of cell discharges in the lateral geniculate nucleus. *Vision Res.,* 11 (Suppl. 3): 253–268, 1971.

50. MONTERO, V. M., ROJAS, A., AND TORREALBA, F. Retinotopic organization of striate and peristriate visual cortex in the albino rat. *Brain Res.,* 53: 197–201, 1973.

51. MURPHY, E. H., AND BERMAN, N. Receptive fields of area 17 neurons in cats and rabbits. *J. Comp. Neurol.,* 188: 401–428, 1979.

52. MURPHY, E. H., AND RAPPAPORT, E. The effect of ablation of the extrastriate visual system on visual reversal learning in the rabbit. *Neurosci. Abstr.,* 3: 570, 1977.

53. NEWSOME, W. T., BAKER, J. F., MEIZEN, F. M., MYERSON, J., PETERSON, S. E., AND ALLMAN, J. M. Functional localization of neuronal response properties of extrastriate visual cortex of the owl monkey. *ARVO Abstract,* 174, 1978.

54. OLAVARRIA, J. A horseradish peroxidase study of the projections from the latero-posterior nucleus to three lateral peristriate areas in the rat. *Brain Res.,* 173: 137–141, 1979.

55. OLAVARRIA, J., AND MENDEZ, B. The representations of the visual field on the posterior cortex of *Octodon degus. Brain Res.,* 161: 539–543, 1979.

56. OLAVARRIA, J., AND MONTERO, V. M. Reciprocal connections between striate cortex and extrastriate cortical visual areas in the rat. *Brain Res.,* 217: 358–363, 1981.

57. OLAVARRIA, J., AND TORREALBA, F. The effect of acute lesions of the striate cortex on the retinotopic organization of the lateral peristriate cortex in the rat. *Brain Res.,* 151: 386–391, 1978.

57a. OLAVARRIA, J., AND VAN SLUYTERS, R. C. An HRP study of the projection from visual cortical areas to the superior colliculus in the rat. *Neurosc. Abstr.* 1981.

58. PALMER, L. A., ROSENQUIST, A. C., AND TUSA, R. J. The retinotopic organization of lateral suprasylvian visual areas in the cat. *J. Comp. Neurol.,* 177: 237–256, 1978.

59. PERRY, V. H. A tectocortical visual pathway in the rat. *Neuroscience,* 5: 915–927, 1980.

60. RIBAK, C. E., AND PETERS, A. An autoradiographic study of the projections from the lateral geniculate body of the rat. *Brain Res.,* 92: 341–368, 1975.

61. ROBSON, J. A., AND HALL, W. C. The organization of the pulvinar in the grey squirrel (*Sciurus carolinensis*). 1. Cytoarchitecture and connections. *J. Comp. Neurol.*, 173: 355–388, 1977.

62. ROCHA-MIRANDA, C. E., LINDEN, R., VOLCHAN, E., LENT, R., AND BOMBARDIERI, R. A., JR. Receptive field properties of single units in the opossum striate cortex. *Brain Res.*, 104: 197–219, 1976.

63. ROJAS, J. A., MONTERO, V. M., AND ROBLES, L. Organización funcional de la corteza visual de la rata. *Proceed. VI Congreso Asoc. Latinoam. Ciencias Fisiol.*, Viña del Mar, Chile, p. 98, 1964.

64. ROSE, M. Cytoarchitektonischer Atlas der Grosshirnrinde des Kaninchens. *J. Psychol. Neurol., Leipzig.*, 43: 354–440, 1931.

65. SANIDES, F., AND HOFFMANN, J. Cyto- and myeloarchitecture of the visual cortex of the cat and of the surrounding integration cortices. *J. Hirnforsch.*, 11: 79–104, 1969.

66. SCHATZ, C. J., LINDSTROM, S. H., AND WIESEL, T. N. The distribution of afferents representing the right and left eyes in cat's visual cortex. *Brain Res.*, 131: 103–116, 1977.

67. SHAW, C., YINON, U., AND AUERBACH, E. Receptive fields and response properties of neurons in the rat visual cortex. *Vision Res.*, 15: 203–208, 1975.

68. SPEAR, P. D., AND BAUMANN, T. P. Receptive field characteristics of single neurons in lateral suprasylvian visual area of the cat. *J. Neurophysiol.*, 38: 1403–1420, 1975.

69. SPEAR, P. D., AND BAUMANN, T. P. Effects of visual cortex removal on receptive field properties of neurons in the lateral suprasylvian visual area in the cat. *J. Neurophysiol.*, 42: 31–56, 1979.

70. STRASCHILL, M., AND HOFFMANN, K. P. Functional aspects of localization in the cat's tectum opticum. *Brain Res.*, 13: 274–283, 1969.

71. THOMPSON, J. M., WOOLSEY, C. N., AND TALBOT, S. A. Visual areas I and II of cerebral cortex of rabbit. *J. Neurophysiol.*, 13: 277–288, 1950.

72. TIAO, Y.-C., AND BLAKEMORE, C. Functional organization in the visual cortex of the golden hamster. *J. Comp. Neurol.*, 168: 459–482, 1976.

73. TORREALBA, F., MONTERO, V. M., AND CARRASCO, M. A. Campos receptivos unitarios en la corteza visual primaria de la rata. *Acta Physiol. Latino Amer.*, 25: 49, 1975.

74. TOWNS, L. C., GIOLLI, R. A., AND HASTE, D. A. Corticocortical fiber connections of the rabbit visual cortex: A fiber degeneration study. *J. Comp. Neurol.*, 173: 537–560, 1977.

75. TURLEJSKI, K., AND MICHALSKI, A. Clare-Bishop area in the cat: Location and retinotopical projection. *Acta Neurobiol. Exp.*, 35: 179–188, 1975.

76. TUSA, R. J., PALMER, L. A., AND ROSENQUIST, A. C. The retinotopic organization of area 17 (striate cortex) in the cat. *J. Comp. Neurol.*, 177: 213–236, 1978.

77. TUSA, R. J., PALMER, L. A., AND ROSENQUIST, A. C. Retinotopic organization of areas 18 and 19 in the cat. *J. Comp. Neurol.*, 185: 657–678, 1979.

78. WAGOR, E., MANGINI, N. J., AND PEARLMAN, A. L. Retinotopic organization of striate and extrastriate visual cortex in the mouse. *J. Comp. Neurol.*, 193: 187–202, 1980.

79. WHITTERIDGE, D. Projection of optic pathways to the visual cortex. In: *Handbook of Sensory Physiology*, edited by R. Jung. Berlin: Springer-Verlag, vol. VII/3, pp. 247-268, 1973.

80. WIENSENFELD, Z., AND KORNEL, E. E. Receptive fields of single cells in the visual cortex of the hooded rat. *Brain Res.*, 9: 401–412, 1975.

81. WOOLSEY, C. N. Comparative studies on cortical representation of vision. *Vision Res.*, 11 (Suppl. 3): 365–382, 1971.

82. WOOLSEY, C. N., SITTHI-AMORN, C., KEESEY, U. T., AND HOLUB, R. A. Cortical visual areas of the rabbit. *Soc. Neurosc. Mtg. San Diego, Abstracts*, p.180, 1973.

83. WOOLSEY, C. N., SITTHI-AMORN, C., KEESEY, U. T., AND HOLUB, R. A. Multiple representations of the visual field in the cerebral cortex. *Scient. Progr. Manuscr., Amer. Ass. Neurol. Surg., St. Louis Meet.*, p. 6, 1974.

84. WOOLSEY, T. A. Somatosensory, auditory and visual cortical areas of the mouse. *Johns Hopkins Med. J.*, 121: 91–112, 1967.

85. ZEKI, S. M. Functional organization of a visual area in the posterior bank of the superior temporal sulcus of the rhesus monkey. *J. Physiol., London*, 236: 549–573, 1974.

86. ZEKI, S. M. The response properties of cells in the middle temporal area (area MT) of owl monkey visual cortex. *Proc. Roy. Soc. London B.*, 207: 329–348, 1980.

Chapter 3

Multiple Representations of the Visual Field

Corticothalamic and Thalamic Organization in the Cat

B. V. Updyke

*Department of Anatomy, LSU Medical Center,
New Orleans, Louisiana*

1. Introduction

In order to understand the functional implications of multiple cortical representations of the visual field, it would seem necessary to first understand the details of their interconnections. This chapter explores one aspect of this approach in the visual system of the cat.

The lateral suprasylvian cortex of the cat has been shown by electrophysiological mapping to contain multiple representations of the visual field, and this mapping evidence, together with evidence for distinct patterns of anatomical connections, has been interpreted to indicate that the lateral suprasylvian cortex contains some ten distinct visual areas, each with a partial representation of the visual hemifield (22, 30). These ten areas exist in addition to visual areas 17, 18 and 19, each of which also contains a single representation of the visual field in the cat (30, 31). Present evidence suggests that most of the lateral suprasylvian visual areas are sparsely interconnected at the cortical level (7,11,27), but are heavily interconnected with the lateral posterior complex of the cat's thalamus (3, 6, 9, 11, 15, 18, 24, 29). To this extent the lateral posterior complex appears to be important for traffic between the various cortical visual areas. Thus, if we are to understand how information is transferred between these cortical areas, it would seem to be useful to determine how the separate areas are interconnected via the thalamus.

It appears to be possible to sort out these interconnections by exploiting what is presently known of the organization of the cat's lateral posterior complex and cortical visual areas. The following sections will discuss evidence for systematic representations of the visual field in the lateral posterior complex and current concepts of its organization, and will consider the applications of that organization to studying the interconnections among the visual areas of the cat's cortex.

2. Retinotopic Organization in Lateral Posterior Complex

Evidence that the cat's lateral posterior complex contains retinotopically organized zones has been obtained primarily from analysis of corticothalamic projections originating from visual areas 17, 18 and 19 (33, 34). These cortical projections were studied autoradiographically (5) following intracortical injections of tritiated amino acid into sites selected on the basis of available maps of the cat's visual areas (21, 26, 30). Because these observations make up a major part of the evidence for systematic visual organization in this part of the cat's thalamus, they are reviewed here in some detail.

2.1. Projections of Area 17

Figure 3.1 shows the characteristic pattern of termination in the thalamus demonstrated by injections of tritiated amino acid into area 17 in the cat. Three main foci of labeled terminals are evident within the lateral geniculate nucleus, medial interlaminar nucleus and lateral posterior complex, and each of the terminations has the

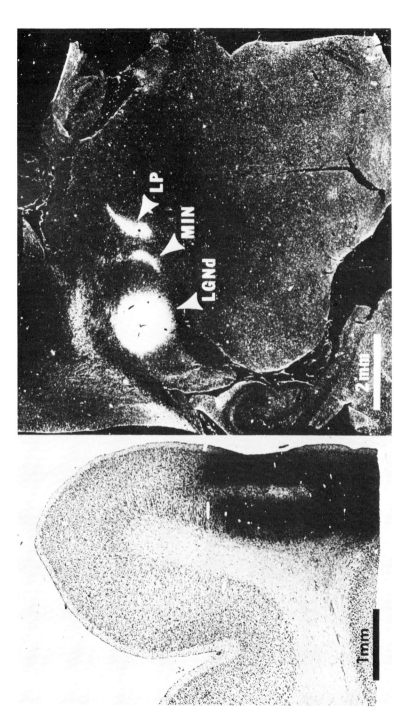

Fig. 3.1. Example of the terminal labeling pattern in lateral thalamus resulting from an injection of tritiated amino acid into area 17 in the cat. A lightfield photomicrograph of the injection site is shown at the left: silver grains overlying the labeled tissue appear black. A dark-field photomicrograph of the termination sites is shown at the right: silver grains overlying labeled terminals appear white. A single injection into area 17 labels axon terminals that occupy prominent bands in the dorsal lateral geniculate nucleus (LGNd), medial interlaminar nucleus (MIN) and the lateral posterior complex (LP).

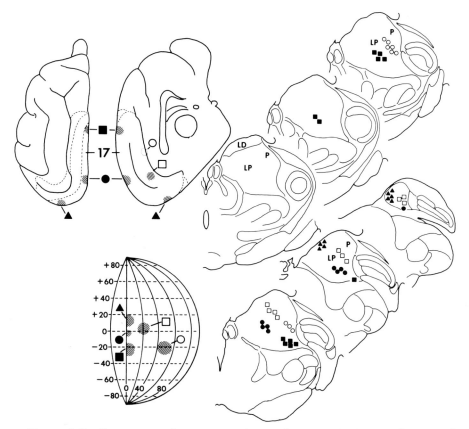

FIG. 3.2. Summary of experiments employing injections of tritiated amino acid into area 17 in the cat. The locations of the injections are shown on the drawings of the dorsal and medial surfaces of the cortex. Estimated visual field loci corresponding to each cortical injection are shown on the perimeter chart. The corresponding terminations in the lateral posterior complex are shown at the right (LD, lateral dorsal nucleus; LP, lateral posterior complex; P, pulvinar). All sites explored project to a narrow zone of the lateral posterior complex. Sites involving the representation of the vertical meridian in area 17 (solid symbols) project in sequence along the medial edge of this zone. Sites involving the representation of the lower quadrant of the visual field project rostrally and those involving the representation of the upper quadrant project caudally. Injection sites, which involved the representation of the lateral periphery of the visual field (open symbols), project near the lateral limit of the termination zone. Results redrawn from Updyke (34).

form of a restricted column or band. Within the dorsal lateral geniculate nucleus and medial intralaminar nucleus, these bands have been shown to align with the lines of retinal projection in the nuclei (32, 34). Within the lateral posterior complex, the bands of label showed an oblique anteroposterior orientation (33, 34).

Since it is reasonable to assume that the area of the visual field represented at the cortical injection site is also represented by the labeled axon terminals, and since each termination takes the form of a restricted band, it is apparent that the cortical topology is preserved in the projections onto the thalamus. Consequently the retinotopic organization of the visual cortex and its projections can be employed as a tool for studying thalamic organization, provided that the visual field loci represented at cortical injection sites can be determined. In the absence of direct mapping evidence, the cortical representation can be estimated from the corticogeniculate projection. Since the projections of area 17 to the lateral geniculate nucleus are retinotopically organized (6, 19), comparison of the position of the corticogeniculate projection with Sanderson's (25) maps of this nucleus provides an indication of the visual field locus represented at the injection site.

This approach was adopted to study the organization of visual cortical projections to the lateral posterior complex in the cat (33, 34). The results of a number of experiments involving area 17 are summarized in Fig. 3.2. All of the sites explored in area 17 project into a narrow zone near the lateral edge of the lateral posterior complex in a pattern that reproduces the cortical organization. Injection sites situated along the representation of the vertical meridian in area 17 project in sequence along the medial margin of the termination zone. Those injection sites that involve the representation of the lower quadrant of the visual hemifield project rostrally and ventrolaterally within this zone, and injection sites that involve the representation of higher elevations in the visual hemifield project progressively more caudally within the zone. Injections that involve the representation of parts of the lateral periphery of the visual hemifield project near the lateral margin of the termination zone. Thus, the pattern that emerged from these experiments suggested the existence of a systematic representation of the visual hemifield within this zone of the lateral posterior complex.

2.2. Projections of Area 18

Experiments of the same type were repeated for cortical area 18. These results are summarized in Fig. 3.3. The projections from area 18 to the lateral posterior complex terminate within the same zone as those from area 17. Again injection sites that involve the cortical representation of the vertical meridian project in rostrocaudal sequence along the medial edge of this zone. Injections that involve the lower quadrant representation project rostrally, and those that involve the representation of higher elevations in the visual hemifield project more caudally. The cortex, representing more

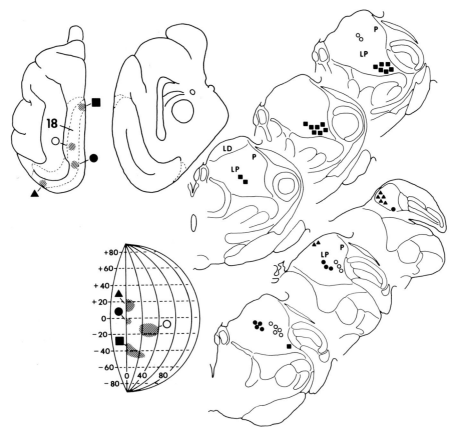

FIG. 3.3. Summary of experiments employing injections of tritiated amino acid into area 18 in the cat. The conventions are the same as those in Fig. 3.2. Projections from area 18 to the lateral posterior complex terminate in register with the projection from area 17. Injection sites that involve the representation of the vertical meridian in area 18 (solid symbols) project along the medial edge of the termination zone. Sites that involve the lower quadrant representation project rostrally and those that involved higher elevations in the visual field representation project more caudally. Cortical sites that represent parts of the lateral periphery of the visual field (open symbols) project near the lateral limit of the termination zone. Results redrawn from Updyke (34).

lateral parts of the visual hemifield in area 18, projects more laterally within the termination zone.

These observations indicate that the projections from area 18 to the lateral posterior complex are also retinotopically organized. When the areas of the visual hemifield represented by the termination patterns are taken into account, it is apparent that the projections of areas 17 and 18 both superimpose in register, defining a single retinotopically organized thalamic zone.

2.3. Projections of Area 19

These observations were confirmed by additional experiments involving area 19. Area 19 also projects upon this same zone and upon a second zone that is partly coextensive with Rioch's (23) pulvinar nucleus. These results are summarized in Fig. 3.4. Each injected locus in area 19 projects to two distinct loci within the lateral

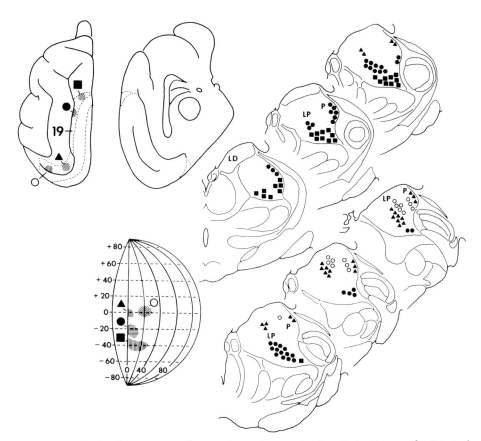

FIG. 3.4. Summary of experiments employing injections of tritiated amino acid into area 19 in the cat. The conventions are the same as those in Fig. 3.2. Each injection site in area 19 projects to two distinct loci within the lateral posterior complex. Injection sites that involve the representation of the vertical merdian (solid symbols), project in sequence along the medial edge of the zone receivng projections from areas 17 and 18 and again along the lateral margin of the pulvinar. Sites involving the representation of the lower quadrant of the visual field project rostrally within both zones and sites involving the representation of higher elevations project more caudally. Cortical sites that represent more lateral parts of the visual field (open symbols) project near the medial limits of pulvinar and the lateral limits of the other zone. Results redrawn from Updyke (34).

posterior complex. Injection sites involving the representation of the vertical meridian in area 19 project in sequence along the medial edge of the zone receiving the projections from areas 17 and 18, and again along the lateral margin of the pulvinar. The injection sites that involve the representation of the lower quadrant of the visual field project rostrally in both termination zones, and those sites involving the representation of higher elevations in the visual field project more caudally. Injections that involve the representation of the lateral periphery of the visual hemifield in area 19 project near the medial limit of the termination zone within pulvinar, and again near the lateral limit of the other zone.

The projections from area 19 thus organize into two distinct retinotopic patterns. One pattern superimposes in register with the projections from areas 17 and 18 and the second mirror image pattern occupies the pulvinar. These results extend earlier observations that the visual cortical projections to the lateral posterior complex in the cat occupy restricted territories (6, 10, 11, 15), and they clear up certain ambiguities about the relative overlap of the separate projections by showing that these projections are related in a systematic way to the representation of the visual hemifield. The projection patterns of areas 17, 18 and 19 thus provide evidence for two distinct retinotopically organized zones within this area of thalamus.

2.4. *Form of the Visual Field Representations*

The resolution of the anatomical techniques employed and the magnification factor distortions of the central visual representations necessarily limit the accuracy with which these thalamic representations of the visual hemifield can be determined. At present, the electrophysiological mapping results from this part of thalamus are limited (16, 17), and more complete mapping data might be expected to reveal the finer details of organization of the visual zones which are not shown by the connectional patterns. A working estimate of the extent and the organization of these two zones can be obtained, however, from the data plotted in Figs. 3.2–3.4. It is apparent that the vertical meridian is represented once along the lateral margin of the pulvinar and again along the medial margin of the second zone. The common boundary of these two zones appears to correspond to some peripheral limit of the visual field representations. Separate loci in the visual hemifield are represented as oblique bands that shift ventrolaterally in anteroposterior sequence. As the result, the lower quadrant of the hemifield is represented rostrally and somewhat ventrolaterally and the upper quadrant is represented caudally and somewhat dorsomedially within each zone.

3. Lateral Posterior Complex Organization

3.1. Subdivisions Based on Connectional Patterns

At the time the observations on these corticothalamic projections were made, it was apparent that the concept of retinotopically organized zones was not easily reconciled with existing cytoarchitectural schemes for dividing the cat's lateral posterior complex (20, 23). To avoid the inconsistencies of having a single retinotopic map occupy parts of several different nuclei, it seemed necessary to attempt to redivide the complex in a manner consistent with its afferent connections (34). It seemed appropriate to distinguish both of the two zones that receive the retinotopically organized cortical projections from a third region that had been shown to receive a projection from the superior colliculus (10, 14). These considerations suggested the scheme shown in Fig. 3.5. It appeared that the entire complex could be divided into four zones—two lateral ones definable by the cortical projections, an intermediate one definable by tectal projections and a medial one lacking visual connections. The term pulvinar was retained for the lateralmost division recipient of the projection from area 19. The remaining zones were designated the lateral, interjacent and medial division of the lateral posterior complex (LPl, LPi, LPm). Zone LPl receives the projections of areas 17 and 18 and zone LPi receives the projections of the superior colliculus.

Although this scheme of divisions must be considered somewhat provisional as long as our knowledge of the connectional patterns is incomplete, the existence of the visual zones is now supported by a number of additional observations on the patterns of cortical and subcortical projections to the cat's lateral posterior complex. That evidence from patterns of cortical projections is discussed in a subsequent section. The subcortical connections which appear to respect the boundaries of the three visual zones include the projections from the superior colliculus to zone LPi (3, 8, 10, 14) and the projections from the pretectal complex to the pulvinar (1, 4, 10). Berson and Graybiel (4) have also recently pointed out that the pulvinar and the tectorecipient zone (LPi), but not zone LPl, stain selectively for acetylcholinesterase. Although the source of this enzyme activity has not yet been definitely established in the cat, it would appear to be associated with yet another system of fibers that has its origin in the midbrain (28).

In keeping with the provisional status of this scheme, other features of the organization of the lateral posterior complex remain unresolved. The pulvinar zone in particular presents some curious

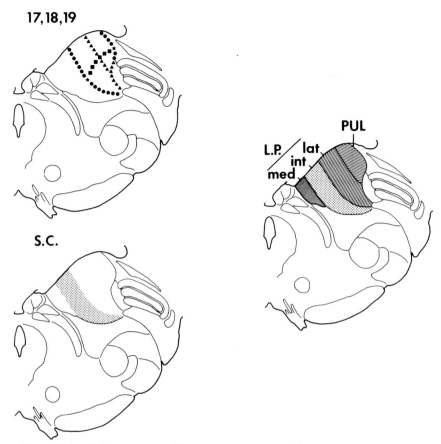

FIG. 3.5. Schematic to illustrate a proposal for dividing the cat's lateral posterior complex into functional zones (34). Two of the divisions of this region can be defined by the projections arising from visual cortical areas 17, 18 and 19 (upper left); a third division can be defined by projections from the superior colliculus (SC, lower left). The scheme of divisions shown at the right recognizes three visual zones: pulvinar (PUL), recipient of projections from area 19; lateral division of the lateral posterior complex (LPl), recipient of projections from areas 17, 18 and 19; interjacent division of the lateral posterior complex (LPi), recipient of projections from superior colliculus. A medial division of the lateral posterior complex (LPm) is not known to receive visually related connections.

features that suggest the possibility of additional subdivisions. Berman and Jones (2) have shown that retinal fibers terminate within a thin interrupted sheet at the extreme lateral margin of the cat's pulvinar, and this same area has been reported to receive a projection from area 17 (3) and from the cerebellum (12). Given that the afferent connections of this marginal zone are different from the remainder of the pulvinar, it may be useful to consider it a distinct subdivision. With respect to the general organization of the cat's lateral posterior complex, Berson and Graybiel (3, 4) have proposed an alternative system of divisions and nomenclature. Their system also recognizes the integrity of the three visually connected zones, but differs from this author's interpretation in recognizing an additional area at the rostral limit of the lateral posterior complex and in lumping the medial part of the complex with the suprageniculate nucleus. A decision on the relative merits of the two alternative interpretations will probably require some additional information about the afferent and efferent connections of the disputed zones.

3.2. Applications to Study of Cortical Interconnections

The unresolved problems of lateral posterior complex organization do not prevent exploiting what is presently known about visual organization in this area to study the interconnections of cortical visual areas. The retinotopically organized zones of the lateral posterior complex appear to be a useful key to cortical organization. These zones appear to be the functional units of organization for this part of thalamus, in much the same way that separate cortical representations of the visual field may be considered functional entities. To the extent that the zones are defined by connectional patterns, they presumably are also the units by which the afferent and efferent connections of the lateral posterior complex are organized. This systematic organization in the thalamus thus greatly simplifies the problems of studying those interconnections between cortical areas that are routed via the thalamus. The identification of retinotopic representations of the visual field in both the lateral posterior complex and in the visual areas of the cat's cortex (22, 30) makes it possible to systematically explore the interconnections between the cortical visual areas and the thalamus at the level of the individual maps.

At present different groups of investigators are only beginning to explore this approach. Some of the findings are reviewed in the following sections. Since this work is in its early stages, the results necessarily represent only a modest beginning toward understanding the connectivities of the cat's cortical visual areas.

4. Interconnections of Cortex and Thalamus

4.1. Corticothalamic Connections

It is known from earlier studies that the cat's lateral suprasylvian cortex projects heavily upon the lateral posterior thalamus (11, 15). These studies predate the identification of systematic representations of the visual field in cortex and thalamus, however, and consequently their findings are now of limited use in addressing the details of organization of these corticothalamic connections.

This investigator has reexamined the projections from the lateral suprasylvian areas with autoradiographic methods (35). In these experiments electrophysiological mapping was used to locate and identify the various lateral suprasylvian areas as described by Tusa et al. (30). The mapping data were also used to estimate the parts of the visual field represented within the cortical areas explored. In general, these mapping results agreed closely with the results reported by Tusa et al. (30) with respect to the representation of the visual field within different areas of the lateral suprasylvian cortex, and I have followed their nomenclature for the various areas. After recording, the explored area was injected with tritiated amino acid and the projections onto the lateral posterior complex were demonstrated autoradiographically.

One of these experiments, involving area 21a is summarized in Fig. 3.6. The projection to the lateral posterior complex in this case takes the form of two separate bands, one within pulvinar and one within zone LPl. This projection pattern is very similar to that obtained from injections of area 19 (34) and, if one compares the position of the labeled terminals in Fig.3.6 with the position of the comparable projection from area 19 (solid triangles, Fig. 3.4), it is clear that the two projections fall in close register. This finding has been consistent for all of the lateral suprasylvian areas explored to date. As with areas 17, 18 and 19, the lateral suprasylvian areas also project retinotopically onto the visual zones of the lateral posterior complex.

Some of the results of this series of experiments are summarized, along with the earlier study of areas 17, 18 and 19 in Table 3.1. These results are arranged to emphasize the finding that there appear to be four separate variations in the pattern of visual cortical projections onto the lateral posterior complex. Areas 17 and 18 project retinotopically upon zone LP1. Areas 19, 21, PMLS and AMLS project retinotopically to both pulvinar and LPl. Areas PLLS and ALLS project only to zone LPi. Area 20 projects to the pulvinar, LPl and LPi. Of the areas not listed in the table, area DLS also appears to project exclusively to zone LPi.

These four projection patterns appear to distinguish between groups of visual areas that occupy different cortical territories. Areas 17 and 18 on the suprasplenial, lateral and posterolateral gyri connect only with zone LPl. The areas on and flanking the suprasylvian gyrus (19, 21, PMLS, AMLS) exhibit a dual projection onto both LPl and pulvinar. Areas flanking the ectosylvian gyrus (PLLS,

ALLS, DLS) project exclusively to zone LPi. Only area 20 at the confluence of the posterior suprasylvian and posterolateral gyri seems to connect across each of the thalamic zones.

Although it is unlikely that this grouping of cortical areas represents any strict processing hierarchy, it is suggestive that certain cortical areas may occupy logically similar levels of processing, at least with respect to the traffic through the lateral posterior complex. This impression is reinforced by the patterns of thalamocortical projections from zones LPi, LPl and pulvinar. These findings are discussed in the following section.

4.2. Thalamocortical Connections

Analysis of the efferent connections of the cat's lateral posterior complex presents more difficulties than does the study of its afferent connections. Even though recent studies employing the autoradiographic and horseradish peroxidase techniques (3, 4, 29) have overcome the problems of involvment to fibers of passage that complicated earlier studies based on degeneration methods (9, 18), certain problems persist. It remains difficult to restrict injections to single zones of the lateral posterior complex and to interpret the boundaries of many of the cortical visual areas without resorting to electrophysiological mapping. These problems are also aggravated to some extent by the mismatches between the extent of the visual field representations in the zones of the lateral posterior complex and in the lateral suprasylvian visual areas. Given that the visual field is systematically represented in both cortex and thalamus, the absence of thalamocortical projections is probably significant only when comparing equivalent parts of the separate visual field representations. All of these problems contribute to the difficulties in studying the efferent connections of these zones and probably account for some of the discrepancies between recent reports (4, 29). In spite of the difficulties, there is a fair amount of agreement among a number of studies concerning the broad patterns of efferent projection of the visual zones of the cat's lateral posterior complex.

Zone LPi appears to have the most restrictive pattern of efferent connections. Degeneration studies have suggested that this part of the lateral posterior complex projects onto the lateral bank and fundus of the middle and posterior suprasylvian sulcus and to the area of confluence of the posterolateral and posterior suprasylvian gyri (9, 18). This pattern of projections is confirmed by more recent autoradiographic and horseradish peroxidase experiments (3, 4, 29). The territory involved in this projection pattern seems to in-

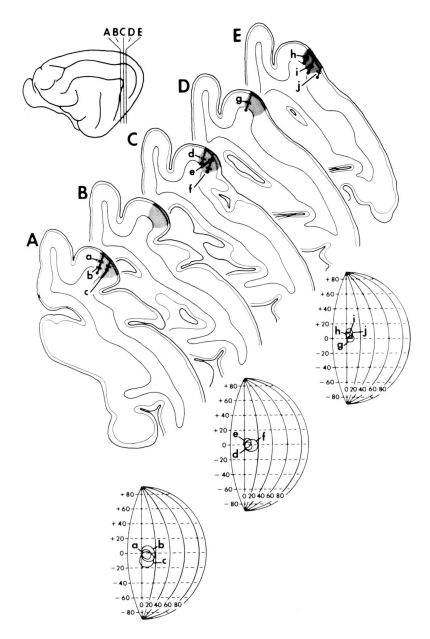

FIG.   3.6. Summary of a combined mapping and injection experiment involving lateral suprasylvian area 21a. A reconstruction of the area explored is shown in the drawings at the left (A–E). The extent of the area labeled by an injection of tritiated amino acid is indicated on the sections by shading. The distribution of receptive fields encountered in this area is

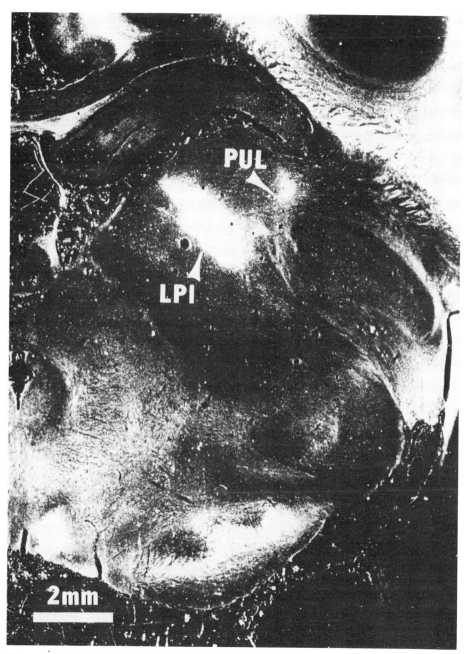

FIG. 3.6. (*continued*)
shown on the perimeter charts (a–j). The projections onto the lateral posterior complex are shown in the dark-field photomicrograph at the right. The projections take the form of two bands, one at the lateral margin of the pulvinar and one at the medial margin of zone LPl. See text for discussion.

Table 3 .1
Projections of Cortical Visual Areas
to Lateral Posterior Complex in Cat[a]

Cortical area	Zones of lateral posterior complex		
	Pulvinar	LPl	LPi
17	−	+	−
18	−	+	−
19	+	+	−
21	+	+	−
PMLS	+	+	−
AMLS	+	+	−
PLLS	−	−	+
ALLS	−	−	+
20	+	+	+

[a]The presence of a projection is indicated by a (+).
Projections of areas 17, 18, 19 and 20 are based on
Updyke (34); those of areas 21, PMLS, AMLS, PLLS and
ALLS are based on Updyke (35).

clude primarily areas PLLS, DLS and 20. The results of Symonds et
al. (29) suggest that areas VLS and 21b might also be involved.

Zone LPl exhibits a much broader pattern of cortical projections. The major projections of this zone distribute to area 19 (3, 4,
18) and to a strip of cortex that includes the medial bank of the middle suprasylvian sulcus, the posteromedial bank of the posterior
suprasylvian sulcus and the adjacent posterior suprasylvian gyrus
(3, 4, 9, 18, 29). This strip thus appears to include areas AMLS,
PMLS, VLS, 20 and 21. Berson and Graybiel (4) report an additional
sparse projection to areas 17 and 18 and Symonds et al. (29) report
that areas PLLS and DLS also receive projections from zone LPl.

By comparison with the projections of LPl, those of the pulvinar
are also restricted. There is reasonable agreement that the pulvinar
projects to area 19, to the splenial visual area of Woolsey (36) and of
Kalia and Whitteridge (13), to area 7 on the crown of the middle
suprasylvian gyrus and to area 20 (3, 4, 18, 29).

Again the efferent connectional patterns tend to emphasize
broad groups of cortical areas that occupy different cortical territories. Zone LPl appears to project primarily to areas flanking the
crown of the middle suprasylvian gyrus and to area 20. Zone LPi
projects primarily to the areas on the lateral bank of the suprasylvian sulcus and to area 20. The pulvinar projects in a scattered
pattern to the crown of the middle suprasylvian gyrus, to the splenial visual area and to area 20.

5. Summary

From the evidence that has accumulated to date, it appears that the connectional patterns of zones LPi, LPl and pulvinar serve not to distinguish separate cortical areas so much as to emphasize broad groupings that occupy comparable positions with respect to their afferent and efferent connections with the thalamus. These thalamic zones would thus appear to relay patterns of information between collections of areas that represent different stages of processing.

Although these projection patterns are suggestive of sequential transfer of information between groups of cortical areas, it may be well to reserve judgement on that point. Much more work needs to be done before a clear picture of cortical connectivity via the thalamus can be obtained from these projection patterns. The connectional patterns of the thalamic zones are characterized both by overlap in the distribution of afferent and efferent connections across zones and by considerable reciprocity in the afferent and efferent connectional patterns of single zones. This overlap and redundancy presents a major obstacle to understanding how information moves across these pathways. It is evident that much further work on these corticothalamocortical pathways, as well as on the direct corticocortical connections, will be required before any clear picture of connectivities between the cat's visual areas will be possible.

Acknowledgment

Supported by Grant EYO-1925 from the National Eye Institute.

References

1. BERMAN, N. Connections of the pretectum in the cat. *J. Comp. Neurol.*, 174: 227–254, 1977.
2. BERMAN, N., AND JONES, E. G. A retinopulvinar projection in the cat. *Brain Res.*, 134: 237–248, 1977.
3. BERSON, D. M., AND GRAYBIEL, A. M. Parallel thalamic zones in the LP-pulvinar complex of the cat identified by their afferent and efferent connections. *Brain Res.*, 147: 139–148, 1978.
4. BERSON, D. M., AND GRAYBIEL, A. M. Thalamocortical projections and histochemical identification of subdivisions of the LP-pulvinar complex in the cat. *Soc. Neurosci. Abstr.*, 4: 620, 1978.

5. COWAN, W. M., GOTTLIEB, D. I., HENDRICKSON, A. E., PRICE, J. L., AND WOOLSEY, T. A. The autoradiographic demonstration of axonal connections in the central nervous system. *Brain Res.*, 37: 21–51, 1972.

6. GAREY, L. J., JONES, E. G., AND POWELL, T. P. S. Interrelationships of striate and extrastriate cortex with primary relay sites of the visual pathway. *J. Neurol. Neurosurg. Psychiat.*, 31: 135–157, 1968.

7. GILBERT, C. D., AND KELLY, J. P. The projections of cells in different layers of the cat's visual cortex. *J. Comp. Neurol.*, 163: 81–106, 1975.

8. GRAHAM, J. An autoradiographic study of the efferent connections of the superior colliculus in the cat. *J. Comp. Neurol.*, 173: 629–254, 1977.

9. GRAYBIEL, A. M. Some ascending connections of the pulvinar and nucleus lateralis posterior of the thalamus of the cat. *Brain Res.*, 44: 99–125, 1972.

10. GRAYBIEL, A. M. Some extrastriate visual pathways in the cat. *Invest. Ophthalmol.*, 11: 322–333, 1972.

11. HEATH, C. J., AND JONES, E. G. The anatomical organization of the suprasylvian gyrus of the cat. *Ergeb. Anat. Entwicklungs Gesch.*, 45: 7–64, 1971.

12. ITOH, K. A cerebello-pulvinar projection in the cat, revealed by anterograde axonal transport of HRP. *Anat. Rec.*, 193: 745, 1979.

13. KALIA, M., AND WHITTERIDGE, D. The visual areas in the splenial sulcus of the cat. *J. Physiol., London*, 232: 275–283, 1973.

14. KAWAMURA, S. Topical organization of the extrastriate visual system in the cat. *Exptl. Neurol.*, 45: 451–461, 1974.

15. KAWAMURA, S., SPRAGUE, J. M., AND NIIMI, K. Corticofugal projections from the visual cortices to the thalamus, pretectum and superior colliculus in the cat. *J. Comp. Neurol.*, 158: 339–362, 1974.

16. KINGSTON, W. J., VADAS, M. A., AND BISHOP, P. O. Multiple projections of the visual field to the medial portion of the dorsal lateral geniculate nucleus and the adjacent nuclei of the thalamus of the cat. *J. Comp. Neurol.*, 136: 295–316, 1969.

17. MASON, R. Functional organization in the cat's pulvinar complex. *Exptl. Brain Res.*, 31: 51–66, 1978.

18. NIIMI, K., KADOTA, M., AND MATSUSHITA, Y. Cortical projections of the pulvinar nuclear group of the thalamus in the cat. *Brain, Behav. Evol.*, 9: 222–257, 1974.

19. NIIMI, K., KAWAMURA, S., AND ISHIMARU, S. Projections of the visual cortex to the lateral geniculate and posterior thalamic nuclei in the cat. *J. Comp. Neurol.*, 143: 279–312, 1971.

20. NIIMI, K., AND KUWAHARA, E. The dorsal thalamus of the cat in comparison with monkey and man. *J. Hirnforsch.*, 14: 303–325, 1973.

21. OTSUKA, R., AND HASSLER, R. Über Aufbau und Gliederung der corticalen Sehsphäre bei der Katze. *Arch. Psychiat. Nervenkrankh.*, 203: 212–234, 1962.

22. PALMER, L. A., ROSENQUIST, A. C., AND TUSA, R. J. The retinotopic organization of lateral suprasylvian visual areas in the cat. *J. Comp. Neurol.*, 177: 237–256, 1978.

23. RIOCH, D. McK. Studies on the diencephalon of carnivora. I. The nuclear configuration of the thalamus, epithalamus and hypothalamus of the dog and cat. *J. Comp. Neurol.*, 49: 1–119, 1929.

24. ROSENQUIST, A. C., PALMER, L. A., EDWARDS, S. B., AND TUSA, R. J. Thalamic efferents to visual cortical areas in the cat. *Soc. Neurosci. Abstr.*, 1: 53, 1975.

25. SANDERSON, K. J. The projection of the visual field to the lateral geniculate and medial interlaminar nuclei in the cat. *J. Comp. Neurol.*, 143: 101–118, 1971.

26. SANIDES, F., AND HOFFMANN, J. Cyto- and myelo-architecture of the visual cortex of the cat and of the surrounding integration cortices. *J. Hirnforsch.*, 11: 79–104, 1969.

27. SHOUMURA, K. Patterns of fibers degeneration in the lateral wall of the suprasylvian gyrus (Clare-Bishop area) following lesions in the visual cortex of cat. *Brain Res.*, 43: 264–267, 1972.

28. SHUTE, C. C. D., AND LEWIS, P. R. The ascending cholinergic reticular system: Neocortical, olfactory and subcortical projections. *Brain.*, 90: 497–520, 1967.

29. SYMONDS, L., ROSENQUIST, A. C., EDWARDS, S. B., AND PALMER, L. S. Thalamic projections to electrophysiologically defined visual areas in the cat. *Soc. Neurosci. Abstr.*, 4: 647, 1978.

30. TUSA, R. J., PALMER, L. A., AND ROSENQUIST, A. C. The retinotopic organization of the visual cortex in the cat. *Soc. Neurosci. Abstr.*, 1: 52, 1975.

31. TUSA, R. J., PALMER, L. A., AND ROSENQUIST, A. C. The retinotopic organization of area 17 (striate cortex) in the cat. *J. Comp. Neurol.*, 177: 213–235, 1978.

32. UPDYKE, B. V. The patterns of projection of cortical areas 17, 18, and 19 onto the laminae of the dorsal lateral geniculate nucleus in the cat. *J. Comp. Neurol.*, 163: 377–396, 1975.

33. UPDYKE, B. V. Retinotopic organization in the pulvinar and lateral posterior complex of the cat. *Anat. Rec.*, 184: 552, 1976.

34. UPDYKE, B. V. Topographic organization of the projections from cortical areas 17, 18, and 19 onto the thalamus, pretectum, and superior colliculus in the cat. *J. Comp. Neurol.*, 173: 81–122, 1977.

35. UPDYKE, B. V. Projections from lateral suprasylvian cortex to the lateral posterior complex in the cat. *Anat. Rec.*, 193: 707–708, 1979.

36. WOOLSEY, C. N. Comparative studies on cortical representation of vision. *Vision Res.* (Suppl. 3), 11: 365–382, 1971.

Chapter 4

Families of Related Cortical Areas in the Extrastriate Visual System
Summary of an Hypothesis

Ann M. Graybiel and David M. Berson

Massachusetts Institute of Technology, Department of Psychology and Brain Science, Cambridge, Massachusetts

1. Multiple Ascending Channels in the Visual System

Microelectrode studies in the last ten years have uncovered an elaborate multiple representation of the visual field in the posterior association cortex of cats and primates. As many as thirteen "visual areas" have been identified (1, 22, 30, 34, 35), each containing a representation of at least part of the visual field, and still other cortical areas, well beyond the visual association cortex even most generously defined, have been shown to contain neurons responsive to visual stimuli (19, 21). There is no longer any doubt that such multiple representation is a general characteristic of the posterior association cortex, for qualitatively comparable organizations have been demonstrated in the auditory and somatic sensory fields (14, 19, 20, 24, 25, 32, 33).

In step with these physiological findings, anatomical work has shown the sensory pathways leading to the neocortex to be themselves far more complex than previously appreciated. In the visual system there are, in addition to the direct retinogeniculate pathway, a number of indirect lines of conduction leading to the thalamus. The retinogeniculate pathway is itself now known to contain at least three functionally distinct channels originating in the X, Y and W ganglion cells of the retina (see ref. 23). Studies of the thalamocortical projection of the lateral geniculate body suggest that these channels remain individually distinct partly by engaging different cortical layers in area 17 (8, 13, 18). There is also some separation by area of termination in the cat, whose lateral geniculate body projects to the extrastriate cortex as well as to area 17. Though less is known in detail about the so-called extrageniculate pathways, prominent pathways have been shown, in a number of species, to arise in the superior colliculus and pretectal region and to ascend to the posterior thalamus (2–5, 9, 31). In the cat, these fiber pathways, and also a descending projection from the striate cortex, terminate in largely nonoverlapping subfields within the nucleus lateralis posterior (LP) and the pulvinar (5, 9). Thus, there are multiple, relatively discrete representations of the visual modality in the extrageniculate thalamus just as there are in the extrastriate cortex.

2. Experimental Questions and an Hypothesis about the Thalamocortical Connections

These findings concerning the thalamic organization focused our work on two main questions. First, we wanted to know whether the multiple representations of the visual field in the extrastriate cortex are systematically related to particular subdivisions of the LP– pulvinar complex of the thalamus and, thus, by implication, to one or more of the extrageniculate pathways ascending to the thalamus. Second, we wanted to learn to what extent the geniculate and extrageniculate channels might converge at the level of the cortex by means of transcortical association projections. We have attempted to answer these questions by studying the connections of the extrastriate cortex with anterograde and retrograde axon-transport techniques (7, 15–17). The experimental findings, though still incomplete, have led to the hypothesis that there are distinct families of extrastriate cortical areas, each set primarily related to a single subdivision of the LP–pulvinar complex (11). Of these family clusters, it is only the set of cortical areas receiving input from the striate-recipient zone of the thalamus that is heavily interconnected with the striate cortex by direct transcortical pathways. No single hierarchy of striate-to-extrastriate pathways, therefore, seems likely to characterize adequately the organization of this large posterior neocortical field. Instead, the extrastriate areas appear to be related to one another and to the striate cortex according to a much more intricate set of rules that takes into account, as one key determinant of transcortical connectivity, the thalamic affiliations of each cortical zone.

3. Identification of Extrageniculate Thalamic Subdivisions by Their Afferent Connections

Figure 4.1 illustrates the LP–pulvinar complex of the cat in a transverse section stained by the Nissl method and shows a summary plot of the distributions of several of its vision-related afferent pathways. To a first approximation, the entire region can be divided, on the basis of these afferents, into a series of parallel strips starting at the lateral periphery with a thin retinorecipient zone and continuing in sequence with pretectorecipient, striate-recipient and tectorecipient zones (5). These subdivisions can be followed in the longitudinal dimension over approximately the same distance as the lateral geniculate body. At the medial edge of this series lies the complex formed by the nucleus lateralis medialis and the suprageniculate nucleus (the LM–Sg complex, ref. 10). This subdivision may be related to the auditory as well as the visual modality and is itself probably divisible into two subzones.

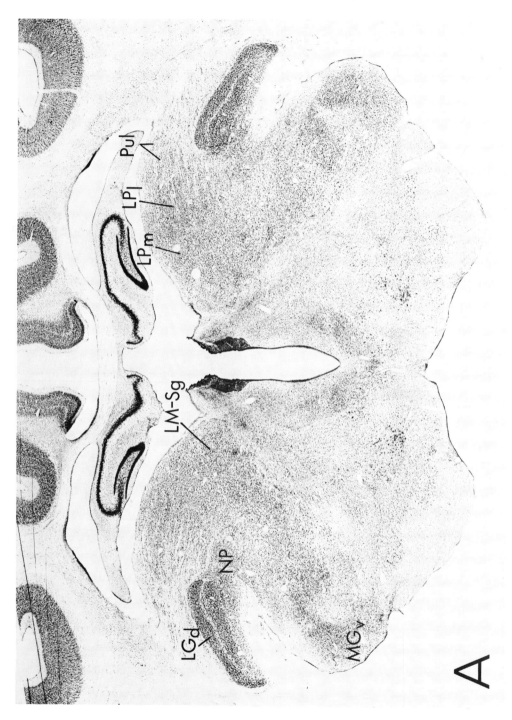

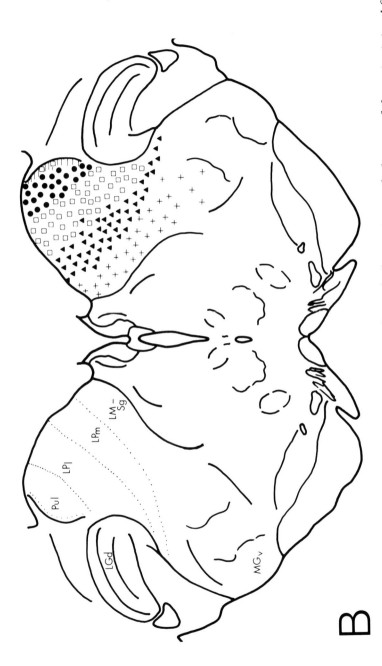

Fig. 4.1. A, photomicrograph of a transverse section through the posterior thalamus of the cat, stained for Nissl substance. B, a schematic drawing illustrating the main afferent-zones of the LP–pulvinar complex and the differential distribution within this complex of afferents from the retina (horizontal lines), pretectum (solid dots), striate cortex (open squares), superior colliculus (solid triangles). Also shown is the adjoining LM–Sg complex with the distribution of afferents from the deep layers of the superior colliculus and insular cortex (cross-marks). Abbreviations: LGd, dorsal lateral geniculate complex: LGv, ventral nucleus of lateral geniculate body: NP, nucleus posterior of Rioch: MGv, ventral subdivision of medial geniculate body: Pul, pulvinar: LP1, lateral division of the nucleus lateralis posterior: LPm, medial division of LP: LM–Sg, complex formed by the nucleus lateralis medialis and suprageniculate nucleus.

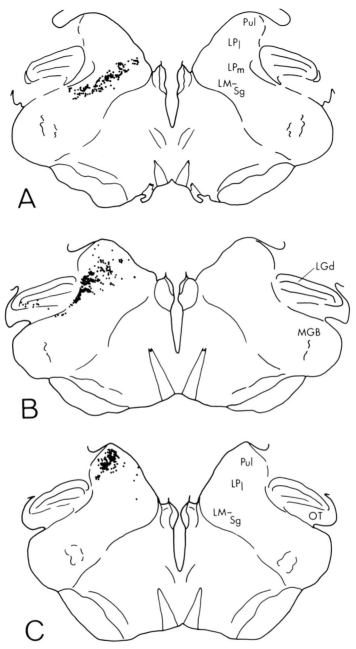

FIG. 4.2. Individual chartings from three cases illustrating strip-like arrangement of neurons labeled by retrograde transport following injections of horseradish peroxidase into the extrastriate cortex of the cat. The particular cortical regions injected were: (A) the lateral bank of the middle suprasylvian sulcus, (B) the medial bank of the middle suprasylvian sulcus and (C) the crown of the middle suprasylvian gyrus. Compare with Fig. 4.1B. Abbreviations as in Fig. 4.1.

4. Evidence for Systematic Groupings of Thalamic–Extrastriate Connections

The crispness with which these extrageniculate subdivisions could be delimited on the basis of their afferent connections encouraged us to study the efferent organization of the LP–pulvinar complex with the retrograde horseradish peroxidase (HRP) cell-labeling technique. It seemed quite likely that we could recognize, at least in outline, how the afferents and efferents were related, because we could inject the tracer substance into one and then another area of the visual cortex, using the maps of Palmer, Tusa and Rosenquist as guides (22, 27–29), and look to see whether the labeled neurons were sprinkled through all the afferent zones or organized in groupings related in some systematic way to these zones, for example, either parallel or perpendicular to them. As shown in Fig. 4.2, the findings clearly indicated that the neurons projecting to a single extrastriate area are grouped together and that the groups tend to have the same strip-like shape, and similar orientations and general locations as the afferent-fiber bands.

Because it is virtually impossible to identify the borders of these afferent subdivisions reliably and consistently in Nissl or myelin stains, we could not make more precise correlations between the afferent and efferent bands simply on the basis of the locations of the labeled neurons. The obvious next step was to carry out double-labeling experiments, tagging a single system of thalamic afferent fibers and the cells of origin of a single thalamocortical projection on the same side of one brain. This turned out to be difficult in practice because an accurate match between labeled afferents and efferents requires experimental marking of all of the constituents of the particular fiber systems in question and no others. We were able in part to circumvent this problem when we learned that the acetylcholinesterase stain could provide a simple and reliable histochemical marker for the subdivisions (10).

Figures 4.3 and 4.4 illustrate this finding. First, as Fig. 4.3 shows, the main subdivisions of the LP–pulvinar complex can be seen at a glance in sections stained by the cholinesterase method. Second, we carried out experiments comparing the subdivisions visible histochemically with patterns of afferent fiber distribution and found, as summarized in Fig. 4.4, that the match between them is quite close (10). In trying to make a coordinated study of the afferents and efferents, we therefore have been able to rely on the chemoarchitecture as a guide, referring patterns of distribution of labeled neurons or fibers to the histochemically distinct subdivisions in the manner illustrated in Fig. 4.4.

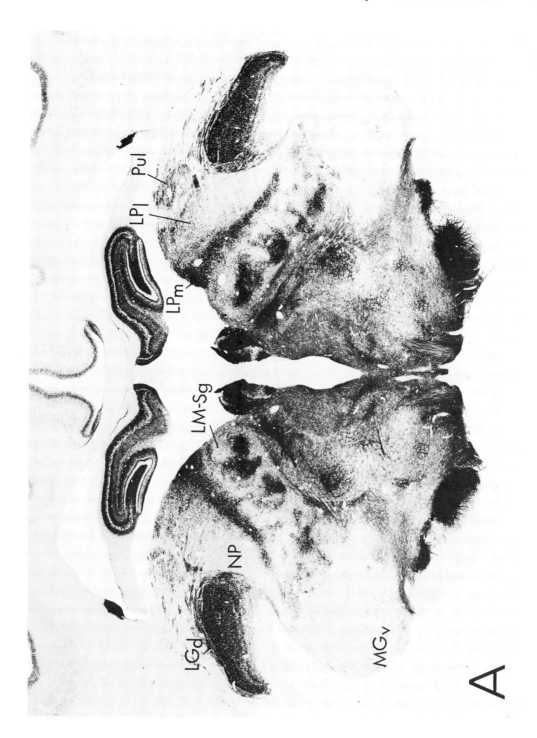

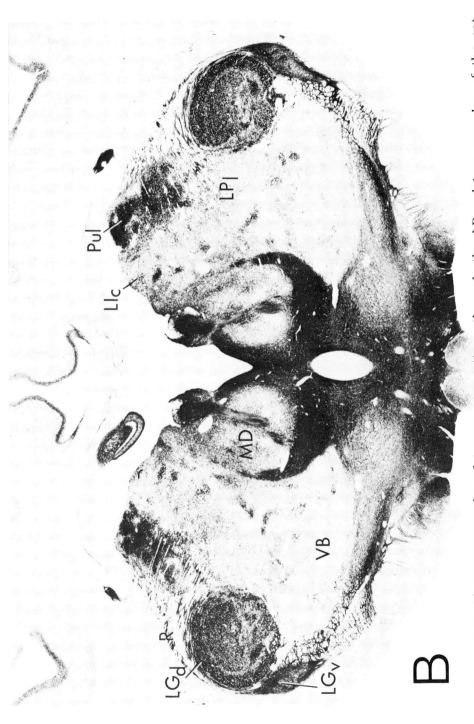

FIG. 4.3. Photomicrographs of transverse sections through the LP–pulvinar complex of the cat illustrating the main subdivisions of the region visible by virtue of their differing contents of the enzyme. acetylthiocholinesterase (10). The more caudal section (A) is the serial neighbor of the Nissl-stained section shown in Fig. 4.1A. Abbreviations as in Fig. 4.1.

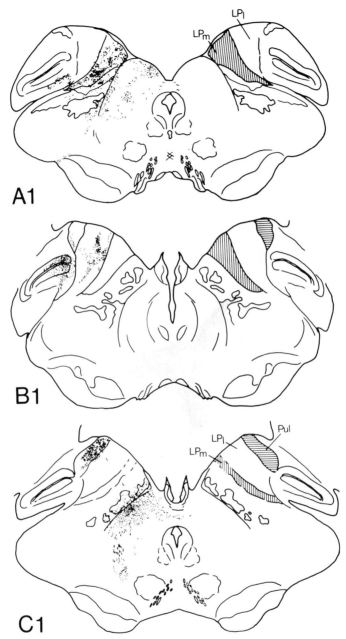

FIG. 4.4. Schematic drawings prepared to illustrate the close match between subdivisions of the LP–pulvinar complex visible in anterograde autoradiographic labeling experiments (A1,B1,C1, left sides), retrograde cell-labeling experiments (A2,B2,C2, left sides) and the acetylcholinesterase stain (right sides of the drawings). A illustrates the tectal projection to LPm and the projection of cells in LPm to the lateral Clare-Bishop

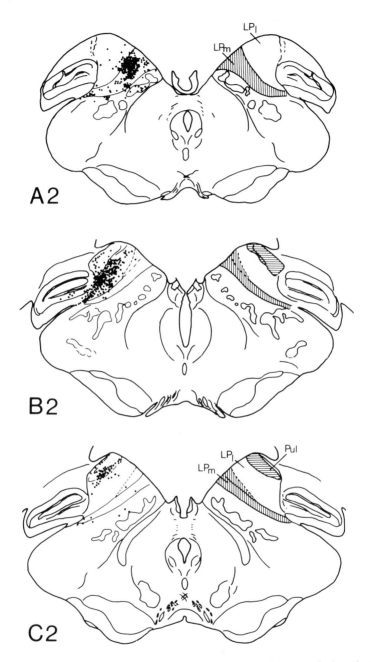

area. B illustrates the projection from area 17 to LP1 and the thalamo-cortical projection from cells in this zone to the medial Clare-Bishop area. C illustrates the distribution of pretectothalamic fibers within the pulvinar and the labeling of neurons in the pulvinar after injections of HRP into the cortex of the middle suprasylvian crown. Compare with Figs. 4.1–4.3.

These new experiments have given us considerably more confidence in drawing conclusions about the through-connectivity. It seems certain, for example, that the cortex of the middle suprasylvian crown receives a thalamocortical projection from the pretectorecipient pulvinar, that the PMLS area of Palmer et al. (22) (corresponding largely to Hubel and Wiesel's Clare-Bishop area; ref. 12) receives fiber projections from the middle part of the LP-pulvinar triad, that is, the striate-recipient LP1 zone, and that the PLLS area (or lateral subdivision of the Clare-Bishop complex) receives thalamic input from the tectorecipient LPm subdivision. Indeed, each of the extrastriate areas we have studied appears to have a primary affiliation with one of the three main extrageniculate subdivisions defined on the basis of its afferent connections. These highly specific thalamocortical relationships are in turn embedded within the framework of a more global topological continuum, so that we cannot be sure of identifying differences between the patterns of connection immediately to either side of a border separating two neighboring zones, either thalamic or cortical. Though we have, as a consequence, followed a method of approximation at these borders, comparing series of experimental findings with one another, the specificity of the relationships illustrated in Fig. 4.4 suggests a dominant plan distinguishing the major thalamocortical connections of each zone of the extrageniculate thalamus.

The second approach we have used to study the thalamocortical connections is aimed at determining the total field of projection of each of the main subdivisions. Figure 4.5 illustrates the results of autoradiographic experiments, in which we attempted to restrict electrophoretic injections of tritiated amino acids to LPm, LP1 and the pulvinar, and also a case of injection in the LM–Sg complex. Though we continue to have problems at the borders, the results strongly support the conclusion that each of these zones has an individually distinct pattern of thalamocortical projection and that each projects to more than one cortical area (see ref. 6, 26). The principal tectorecipient zone (LPm) projects densely to the lateral part of the Clare-Bishop complex (ALLS, PLLS and DLS of Palmer and his colleagues), and either the tectorecipient LPm zone proper or the immediately adjacent part of the LM–Sg complex projects to part of the anterior ectosylvian sulcal cortex. The striate recipient zone (LP1) projects to a separate set of extrastriate areas including the medial part of the Clare-Bishop complex (the medial set of LS areas of Palmer et al.), areas 18, 19, 20 and 21, the splenial visual area and also to area 17 itself. Finally, the pretectorecipient zone (the pulvinar) projects to the crown of the middle suprasylvian gyrus, to ventral splenial cortex, to part of area 20 and sparsely to

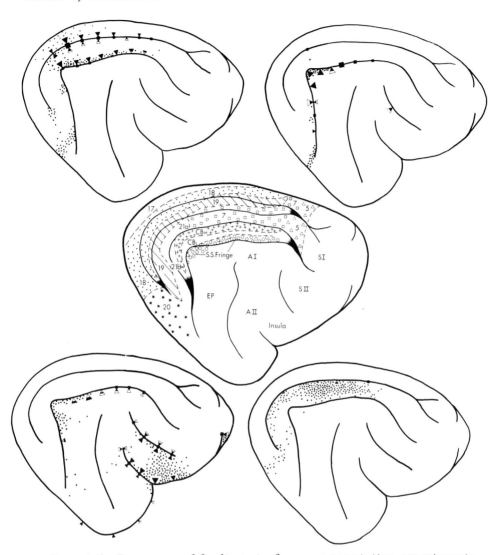

FIG. 4.5. Summary of findings in four representative experiments designed to delimit the total thalamocortical projections of LP1 (upper left), LPm (upper right), the pulvinar (lower right) and the LM–Sg complex (lower left). For each case, the autoradiographic labeling of the cortex, resulting from a circumscribed thalamic injection, has been reconstructed on a schematic lateral view of the hemisphere. Labeling of the crown of a gyrus is shown by dots. Squares and triangles depict, respectively, labeling of the fundus and banks of sulci; size of these symbols indicates density of labeling. Blackening of arrows indicates approximate location of the sulcal labeling along the banks. The central diagram illustrates subdivisions of the cat's extrastriate cortex as determined mainly by the studies of Tusa, Palmer and Rosenquist (22,27,29).

area 19 and the adjoining dorsal part of area 21. There are some regions of apparent overlap. For example, area 19, though receiving dense fiber projections from the striate-recipient LP1, also receives at least a sparse projection from the pretectorecipient pulvinar. Surprisingly few instances of such convergence have appeared, however, and this has led us to conclude that the extrageniculate channels identified at the thalamic level are continued through to the neocortex, so that each channel is related to a particular cluster of extrastriate areas (11).

5. Family Clusters in the Extrastriate Cortex

The pattern of thalamocortical connections just reviewed gives a very different picture of the organization of visual cortex than the classical one, in which a single "core" area (the striate cortex) lies at the center of a single "belt" of extrastriate cortex. Instead, there appear to be several cores, or at least several family clusters, in the visual cortex. Just as the principal geniculostriate pathway defines the group of cortical areas related to the retinogeniculate pathway, so other family clusters are characterized by a preferential relation to the tectothalamocortical and pretectothalamocortical channels and to the transthalamic circuitry of the striate-recipient zone.

To explore this idea further, we have begun to study the transcortical association connections of several of the extrastriate areas. These experiments are far from complete, but the preliminary findings, together with recent observations from other laboratories, suggest that the corticocortical ties linking different extrastriate families are sharply limited, whereas direct transcortical connections are more common for members of a single family. The second pattern to emerge is that only a restricted set of extrastriate areas receives direct transcortical projections from area 17, and these striate-recipient areas of the cortex are, with notably few exceptions, the same extrastriate areas receiving thalamic input from the striate-recipient zone of the thalamic LP–pulvinar complex (Fig. 4.6). It therefore seems likely that the corticothalamocortical and corticocortical circuits emerging from a given part of the neocortex are systematically related and, accordingly, that working out the logic relating transcortical and transthalamic circuits is approachable at least for areas of the extrastriate cortex (11).

It should be possible to carry out a similar coordinated analysis of thalamic and cortical connections in the auditory and somatic sensory areas but we are still at a loss, as the pathways lead farther away from an immediate relation to one of the sensory systems, and

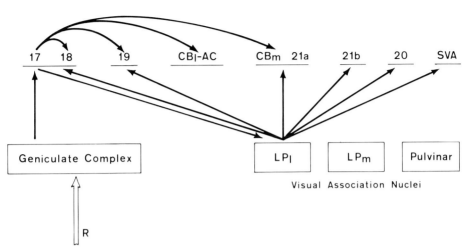

F<small>IG</small>. 4.6. Schematic diagram illustrating the close correspondence between the transcortical projections of area 17 and the thalamocortical projections of the striate-recipient zone of the LP–pulvinar complex (LP1). Cortical areas are shown at the top.

as an increasing proportion of the thalamocortical circuitry is related to the forebrain alone rather than to ascending lines of sensory input. It can be expected that major efforts will be needed in studying these more distal circuitries before we can arrive at a realistic view of the complexity of forebrain connections, and before we can learn whether there are individually distinct access routes leading out from the family clusters of the sensory association cortex toward the effector mechanisms of the forebrain and brain stem. This information will surely be crucial to the enterprise of relating individual circuits to particular functional categories and to the task of finding clues to structure–function relationships in comparative studies of these sensory pathways. Both of these problems have been central to Dr. Woolsey's own work and to the school he founded, and it is a pleasure to contribute this interim review of some of the anatomical connections in honor of his pioneering accomplishments.

Acknowledgments

The studies described here were funded by grants from the National Institutes of Health (R01 EY 02866091 and 1-P30-EY02621) and the National Science Foundation (BNS75-18758 and 78-10549). It is a pleasure to thank Mr. Henry F. Hall and Miss Elaine Yoneoka for their expert technical support.

References

1. ALLMAN, J. M., AND KAAS, J. H. The dorsomedial cortical visual area: a third tier area in the occipital lobe of the owl monkey (*Aotus trivirgatus*). *Brain Res.*, 100: 473–487, 1975.

2. BENEVENTO, L. A., AND EBNER, F. F. The contribution of the dorsal lateral geniculate nucleus to the total pattern of thalamic terminations in striate cortex of the Virginia opossum. *J. Comp. Neurol.*, 143: 243–260, 1971.

3. BENEVENTO, L. A., AND FALLON, J. H. The ascending projections of the superior colliculus in the rhesus monkey (*Macaca mulatta*). *J. Comp. Neurol.*, 160: 339–362, 1975.

4. BENEVENTO, L. A., REZAK, M., AND SANTOS-ANDERSON, R. An autoradiographic study of the projections of the pretectum in the rhesus monkey (*Macaca mulatta*): evidence for sensorimotor links to the thalamus and oculomotor nuclei. *Brain Res.*, 127: 197–218, 1977.

5. BERSON, D. M., AND GRAYBIEL, A. M. Parallel thalamic zones in the LP–pulvinar complex of the cat identified by their afferent and efferent connections. *Brain Res.*, 147: 139–148, 1978a.

6. BERSON, D. M., AND GRAYBIEL, A. M. Thalamo-cortical projections and histochemical identification of subdivisions of the LP–pulvinar complex in the cat. *Soc. Neuroscience Abstr.*, 4: 620, 1978b.

7. COWAN, W. M., GOTTLIEB, D. I., HENDRICKSON, A. E., PRICE, J. L., AND WOOLSEY, T. A. The autoradiographic demonstration of axonal connections in the central nervous system. *Brain Res.*, 37: 21–51, 1972.

8. FERSTER, D., AND LE VAY, S. The axonal arborizations of lateral geniculate neurons in the striate cortex of the cat. *J. Comp. Neurol.*, 182: 923–944, 1978.

9. GRAYBIEL, A. M. Some extrageniculate visual pathways in the cat. *Invest. Ophthal.*, 11: 322–332, 1972.

10. GRAYBIEL, A. M., AND BERSON, D. M. Histochemical identification and afferent connections of subdivisions in the LP–pulvinar complex and related nuclei in the cat. *Neuroscience*, 5: 1175–1238.

11. GRAYBIEL, A. M., AND BERSON, D. M. On the relation between transthalamic and trans-cortical pathways in the visual system. In: *Cerebral Cortex Colloquium*, Cambridge: MIT Press, 1981, pp. 286–319.

12. HUBEL, D. H., AND WIESEL, T. N. Visual area of the lateral suprasylvian gyrus (Clare-Bishop area) of the cat. *J. Physiol. London*, 202: 251–260, 1969.

13. HUBEL, D. H., AND WIESEL, T. N. Laminar and columnar distribution of geniculo-cortical fibers in the macaque monkey. *J. Comp. Neurol.*, 146: 421–450, 1972.

14. KAAS, J. H., NELSON, R. J., SUR, M., LIN, C. S., AND MERZENICH, M. M. Multiple representations of the body within the primary somatosensory cortex of primates. *Science*, 204: 521–523, 1979.

15. KRISTENSSON, K., AND OLSSON, Y. Retrograde axonal transport of protein. *Brain Res.*, 29: 363–365, 1971.

16. LASEK, R., JOSEPH, B. S., AND WHITLOCK, D. G. Evaluation of a radioautographic neuronanatomical tracing method. *Brain Res.*, 8: 319–336, 1968.

17. LA VAIL, J. H., AND LA VAIL, M. M. Retrograde axonal transport in the central nervous system. *Science*, 176: 1416–1417, 1972.

18. LE VAY, S., AND GILBERT, C. D. Laminar patterns of geniculocortical projection in the cat. *Brain Res.*, 113: 1–19, 1976.

19. LOE, P. R., AND BENEVENTO, L. A. Auditory-visual interaction in single units in the orbito-insular cortex of the cat. *Electroenceph. Clin. Neurophysiol.*, 26: 395–398, 1969.

20. MERZENICH, M. M., COLWELL, S. A., AND ANDERSEN, R. A. This volume, Chapter 18.

21. MOHLER, C. W., GOLDBERG, M. E., AND WURTZ, R. H. Visual receptive fields of frontal eye field neurons. *Brain Res.*, 61: 385–389, 1973.

22. PALMER, L. A., ROSENQUIST, A. C., AND TUSA, R. J. The retinotopic organization of lateral suprasylvian visual areas in the cat. *J. Comp. Neurol.*, 177: 237–256, 1978.

23. RODIECK, R. W. Visual pathways. *Ann. Rev. Neurosci.*, 2: 193–225, 1979.

24. ROSE, J. E., AND WOOLSEY, C. N. Cortical connections and functional organization of the thalamic auditory system of the cat. In: *The Biological and Biochemical Bases of Behavior*, edited by H. F. HARLOW AND C. N. WOOLSEY. Madison: University of Wisconsin Press, 1958, pp. 127-150.

25. SUGA, N. This volume, Chapter 22.

26. SYMONDS, L., ROSENQUIST, A., EDWARDS, S., AND PALMER, L. Thalamic projections to electrophysiologically defined visual areas in the cat. *Soc. Neuroscience Abstr.* 4: 647, 1978.

27. TUSA, R. J., PALMER, L. A., AND ROSENQUIST, A. C. The retinotopic organization of the visual cortex in the cat. *Soc. Neurosci. Abstr.*, 1: 52, 1975.

28. TUSA, R. J., PALMER, L. A., AND ROSENQUIST, A. C. The retinotopic organization of area 17 (striate cortex) in the cat. *J. Comp. Neurol.*, 177: 213–236, 1978.

29. TUSA, R. J., ROSENQUIST, A. C., AND PALMER, L. A. Retinotopic organization of areas 18 and 19 in the cat. *J. Comp. Neurol.*, 185: 657–678, 1979.

30. VAN ESSEN, D. C. Visual areas of the mammalian cerebral cortex. *Ann. Rev. Neurosci.*, 2: 227–263, 1979.

31. WEBER, J. T., AND HARTING, J. K. On the connections of the pretectum in the tree shrew (*Tupaia glis*). *Soc. Neurosci. Abstr.*, 1: 45, 1975.

32. WOOLSEY, C. N. Patterns of sensory representation in the cerebral cortex. *Fed. Proc.* 6: 437–441, 1947.

33. WOOLSEY, C. N. Organization of somatic sensory and motor areas of the cerebral cortex. In: *The Biological and Biochemical Bases of Behavior*, edited by H. F. HARLOW AND C. N. WOOLSEY. Madison: University of Wisconsin Press, 1958, pp. 63–82.

34. YIN, T. C. T., AND MOUNTCASTLE, V. B. Visual input to the visuomotor mechanisms of the monkey's parietal lobe. *Science*, 197: 1381–1383, 1977.

35. ZEKI, S. M. Functional specialisation in the visual cortex of the rhesus monkey. *Nature*, 274: 423–428, 1978.

Chapter 5

Cortical and Subcortical Connections of Visual Cortex in Primates

Rosalyn E. Weller[1] and Jon H. Kaas[1,2]

Departments of Psychology[1] and Anatomy,[2]
Vanderbilt University, Nashville, Tennessee

1. Introduction

Over the last several years, considerable progress has been made in understanding the connections of the visual system in primates. Much of this progress has been the result of applications of the relatively new and powerful autoradiographic and histochemical tracing methods that reveal connections in great detail and clarity. The major limitation on further understanding of visual system connections does not seem to be technical at this time, but rather it is our incomplete knowledge of the functional subdivisions of the primate visual system. It is difficult to study connections of parts of the brain for which the organization and divisions into separate areas or nuclei are still unclear. Quite different schemes of cortical organization have been proposed for visual cortex of New World (5–7) and Old World monkeys (96, 111, 116–118) and neither of these schemes includes all of visually responsive cortex. For example, neither organization deals with subdivisions of inferotemporal cortex, a visually responsive region of cortex that in Old World monkeys has received detailed attention (33, 34, 59), without completely resolving the issue of number and boundaries of subdivisions. Efforts have been made to show similarites in cortical organization between Old and New World primates (1, 95, 99), but how similar or dissimilar the two groups of primates are presently remains uncertain. As a further problem in the study of connections of the primate visual system, the organization of visual cortex in an important group, the prosimian primates, is poorly understood.

Because of the problems discussed above, there are clear limits to a productive discussion of visual system connections in primates at this time. It is obvious that it is potentially more valuable to discuss the connections of those visual areas that have been identified across primates over others. Since there is evidence for the existence of three visual areas, V I, V II and MT, in a wide range of primates, and these areas may exist in all primates including humans, understanding their connections is presently the most useful. Other visual areas have been identified only in New World or only in Old World monkeys. Because most of the subdivisions of visual cortex in the New World primate, the owl monkey (2–7), have been based on multiple criteria of electrophysiological evidence of systematic retinotopic maps, clear architectonic features and boundaries, and distinct patterns of connections, we feel the proposed scheme of organization is more probable and less subject to other interpretations than that presently proposed for the Old World monkeys. Therefore, our description of connections of areas other than V I, V II and MT is largely limited to the visual areas defined in the owl monkey. Though the connections of visual areas in the owl monkey are emphasized, it is still not certain whether many of these connections, as well as the visual areas themselves, are present in other primates. Yet, even if some of the details of connections in owl monkeys are not found in other primates, they do provide an important indication of how complex the connection patterns are likely to be in any primate.

Descriptions in this review begin with the major afferent pathways from retina to cortex, followed by the connections between visual cortical areas, including callosal projections, and the projections of subdivisions of visual cortex to subcortical structures. Other topics include the laminar organization and the reciprocity of connections.

2. The Principal Afferent Pathways To Cortex

The retinal ganglion cells of primates project to many subcortical structures, including the suprachiasmatic nucleus of the hypothalamus, the pregeniculate nucleus, the lateral and dorsal terminal nuclei of the accessory optic system and the pretectal and olivary nuclei of the pretectum (see ref. 87 for review). However, only the inputs to the dorsal lateral geniculate nucleus and the superior colliculus are likely to be significant sources of information for cortical processing. The lateral geniculate nucleus is a relay nucleus in which practically all the cells send their axons to striate cortex (63). The retinotectal information reaches cortex through the pulvinar complex, which projects primarily to extrastriate cortex.

All or nearly all retinal ganglion cells in primates project to the lateral geniculate nucleus (16), so that terminations in other structures are presumably either from axon collaterals of the retinogeniculate cells or a small population of ganglion cells without the geniculate target. Retinal ganglion cells have been physiologically classified into three distinct classes and each may have its own separate pattern of central projections. The classification scheme of X, Y and W cells, originally proposed for the cat visual system, seems to be validly applicable to the visual system of primates (27, 71, 73, 74, 77). Most of the primate ganglion cells are X cells. These cells show a sustained discharge to a steady stimulus, have small receptive fields, are color opponent and conduct impulses at moderate velocities, among other characteristics. Such neurons would be useful in detailed form vision. The primate Y cells make up a smaller population of ganglion cells that respond briefly to a change in a steady stimulus, have larger receptive fields, show a broad-band responsiveness to color and conduct impulses rapidly. They most probably relate to the detection of sudden changes in the visual environment. The primate W cells represent a poorly understood small population of remaining neurons that has been difficult to characterize in terms of responses to visual stimuli, and this class of cells undoubtedly includes several functional types. Their role in vision is unclear, but the slow conduction velocities suggest a different functional role for W cells than for the X or Y cells. These separate X, Y, and W systems appear to relay to cortex over largely parallel pathways.

2.1. The Geniculostriate System

The retinal projections to the primate lateral geniculate nucleus are segregated into layers according to the eye of origin and the differ-

ent functional classes of inputs. The basic primate pattern of geniculate lamination consists of two ventral large cell or magnocellular
layers and two dorsal small cell or parvocellular layers (45, 47). Each
individual layer of the two sets of layers receives input from one or
the other eye. Prosimians have two additional layers of quite small
cells, the koniocellular layers, and many primates also have two
thin ventrally located superficial layers of cells of mixed sizes. Both
of these additional sets of layers consist of one layer of contralateral
retinal input and one of ipsilateral retinal input. In addition, the
parvocellular layers of monkeys and hominoids subdivide to
various extents into interdigitating sublayers or leaflets. Finally,
scattered to substantial numbers of neurons are located in the
interlaminar zones between layers and some retinal terminations
are to these sites.

Besides input from one or the other eye being segregated into
geniculate laminae, different classes of ganglion cells also project to
distinct layers in primates. The functional organization of the
geniculostriate relay system is shown in Fig. 5.1. Electrophysiological evidence has indicated that Y cell inputs activate the magnocellular layers and X cell inputs the parvocellular layers (27, 74, 77).
The magnocellular and parvocellular layers contain similar Y and X
cells, respectively, and relay to separate levels or sublaminae of
layers IV and III of striate cortex in macaque monkeys (41; see 38 for
review). No W cell input to the lateral geniculate nucleus has been
identified in either the magnocellular or parvocellular layers, but W
cells may terminate in the unexplored superficial layers and interlaminar zones of monkeys and also in the koniocellular layers of
prosimian primates. The interlaminar regions and koniocellular
layers of galagos have been found to project to layer I of striate cortex (20), which supports the concept of a third relay system to cortex. Thus, there may be largely parallel X, Y and W cell pathways in
primates from the retina to striate cortex.

Although there have been contrary suggestions, geniculate
projections appear to be confined to striate cortex in primates. An
interesting point derived from knowing the different pathways of
the different classes of cells is that, since the projections of the retinal X cells are completely or almost completely restricted to the lateral geniculate nucleus (16, 56, 73), lesions of striate cortex eventually abolish the X cell system by resulting in a transneuronal loss of
X cells from the retina (101). However, because Y and W cells project
to the superior colliculus (56, 73), probably by sustaining collaterals (101), such systems would not be as severely affected by striate
cortical lesions.

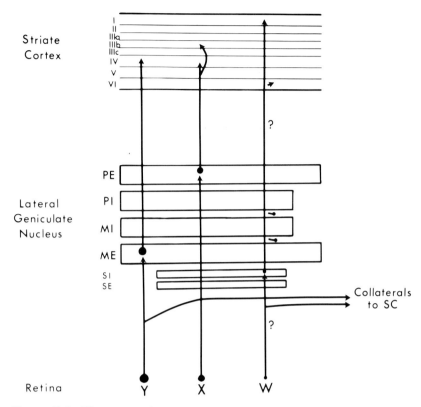

Fig. 5.1 The geniculostriate relay system. Three parallel pathways to cortex relate to X, Y and W retinal ganglion cells. The X and Y pathways have been established by both anatomical and electrophysiological studies in Old World (27, 38, 41, 74) and New World monkeys (77). The W cell pathway proposed here is hypothesized from indirect evidence (see text). Striate cortex layers are after Hassler (37), rather than Brodmann (15). Internal and external parvocellular (PI and PE), magnocellular (MI and ME) and superficial (SI and SE) layers are after Kaas et al. (47). Superior colliculus (SC).

2.2. The Tectopulvinar System

Although a tectopulvinar relay of visual information to cortex has long been known (26, 35), its significance is unclear and we now have only a limited concept of its organization. This second major visual afferent pathway to cortex is shown in Fig. 5.2. This summary is largely based on results from studies on the owl monkey, where details of this system are best known.

The retinal projections to the superficial layers of the superior

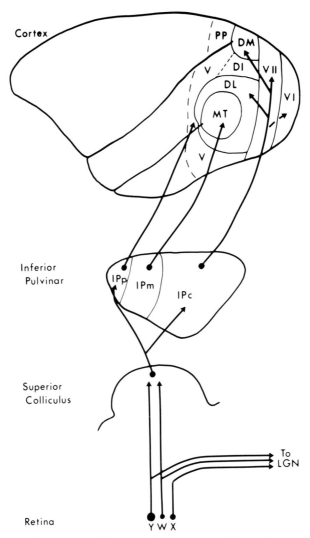

FIG.   5.2.   The tectopulvinar relay system. Retinal input to the superior colliculus from Y and W cells is known from electrophysiological studies in macaque monkeys (56, 73). Studies in owl monkeys (51, 52) indicate that the superior colliculus projects to two of the three subdivisions of
the inferior pulvinar complex, and that each subdivision of the inferior
pulvinar projects to separate regions of extrastriate cortex. The posterior
(IPp), medial (IPm) and central (IPc) nuclei of the inferior pulvinar are from
Lin and Kaas (51). The subdivisions of visual cortex of the owl monkey are
from Allman and Kaas (2–6). Areas V I (primary visual cortex), V II (secondary visual cortex), MT (middle temporal visual area), DL (dorsolateral visual area), and DM (dorsomedial visual area) each contain a topographic
representation of the contralateral visual hemifield and have distinctive
architectonic features. Areas PP (posterior parietal cortex) and DI (dorso-

colliculus in monkeys are evidently mainly from retinal Y cells, with additional input from some of the heterogeneous class of W cells; there is no evidence for input from X ganglion cells (56, 73). Although distinct functional classes of cells are segregated into different layers in the lateral geniculate nucleus, there has been no reported differential distribution of Y and W cells in the primate superior colliculus.

In New World monkeys, such as the owl monkey (51) and the squirrel monkey (Harting, personal communication), the superior colliculus projects to two of the three subdivisions of the inferior pulvinar complex, the posterior, IPp, and the central, IPc, nuclei, but not significantly to the medial nucleus, IPm [after the nomenclature of Lin and Kaas (51)]. In Old World and prosimian primates, subdivisions of the inferior pulvinar have not been established, so the superior colliculus has generally been described as projecting to the "inferior pulvinar" as a whole. However, Lin and Kaas (51) have reviewed evidence suggesting that the superior colliculus also projects to two separate subdivisions of the inferior pulvinar in these primates (e.g., see 11).

For owl monkeys, the relay of retinal information from IPc is primarily but not exclusively to extrastriate cortex, whereas IPp appears to project to visually responsive but unmapped regions of the temporal lobe (51). These tectopulvinar-extrastriate pathways, as well as the pathway from the nontecto-recipient medial subdivision of the inferior pulvinar, IPm (52), to the Middle Temporal Visual Area (MT), are shown in Fig. 5.2. At least part of the inferior pulvinar complex of macaque monkeys projects to large areas of extrastriate cortex, as well as more sparsely to striate cortex (12, 13, 66, 92). However, the details of such relations are not yet clear in Old World primates.

3. A Brief Outline of the Subdivisions of Visual Cortex in Primates

A description of the connections of visual cortex depends on some understanding of the functional subdivisions of visual cortex. Figures 5.2 and 5.3 show some of the subdivisions of visual cortex in owl monkeys detemined from electrophysiological mapping investigations in conjunction with studies of cortical architecture. The

FIG. 5.2 (*continued*): intermediate visual area) are visually responsive, but their topography has not been fully determined (1, 43). The rostral dashed lines mark the extent of visually responsive cortex (V), which includes subdivisions not yet fully defined. Other cortical visual areas of the owl monkey are shown in Fig. 5.3. Lateral geniculate nucleus (LGN).

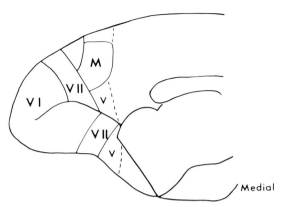

FIG. 5.3. Cortical visual areas of the medial wall of the cerebral hemisphere. A medial view of posterior cortex of the owl monkey. V I, V II and area M (medial visual area) each contain a topographic representation of visual space (7); other conventions as in Fig. 5.2.

justification for these subdivisions is reviewed in this volume by Allman (1) and elsewhere (7, 43). Although the summary figures are detailed, they are still incomplete and much of the organization of visually responsive cortex remains unknown for the owl monkey. In particular, visually responsive cortex rostral to the known areas and other regions, such as the Dorsointermediate Visual Area and the inferotemporal region, are poorly understood.

A quite different scheme of extrastriate cortical organization has been proposed for Old World monkeys (95, 96, 111, 116–118). However, study of these regions in Old World monkeys is more difficult because much of the cortex is buried in fissures and architectonic borders have traditionally been vaguely and inconsistently described. In addition, detailed electrophysiological maps of any regions of the occipital lobe in Old World monkeys have not yet been published. For these reasons, hypotheses of organization in Old World primates have been strongly based on studies of connections and appear to us less certain. Thus, differences between Old and New World monkey subdivisions of extrastriate cortex may be shown by further study to be less pronounced than they seem now.

Only three visual cortical areas have been identified with some degree of certainty in prosimians, New World and Old World monkeys. These visual areas are V I, V II and MT (see Fig. 5.4). Although V I and V II have been considered homologous in all primates, the recognition that MT is the same visual area in New World and Old World primates is very recent (28, 94, 95, 99). Because V I, V II and MT exist in basically the same form in a range of primates, it is reasonable to conclude that they exist in all primates, including humans. Understanding the connections of these areas in any primate would, thus, be most useful in generalizing across primates, and these connections will be described mainly for the owl monkey. A more complete understanding of visual cortex in primates will depend on gathering more comparative data. Although we do not know yet how general all of the owl monkey cortical visual areas are, study of their connections serves to indicate the level of complexity we should expect in the visual system of an advanced primate.

4. Projections of V I

A major source of visual information to extrastriate cortex is from the cortical projections of striate cortex. In all primates that have been studied, striate cortex projects to both V II (we use the term "V II" since it has been consistently defined across species, whereas the term "area 18" has not) and to MT. Projections to these two visual areas have been demonstrated in the prosimian galago (86, 89), a number of New World monkeys (57, 79, 80, 85, 88, 108, 109) and in macaque monkeys (23, 49, 76, 99, 102, 109, 111, 113, 115, 116). As for all other visual system connections studied in detail, the projections of V I to both V II and MT are homotopic. However, the projections from V I to V II are more complicated than those from V I to MT. Although a point in V I projects to a single small area in MT, this is not always true for projections to V II. Though both V I and MT are simple topological representations of the contralateral visual field, V II is what has been described as a "split" or second-order representation in which most of the upper and lower visual fields are separate (4). For this reason, locations along much of the representation of the horizontal meridian in V I project to two separate locations in the two split parts of V II. The topographical relations of the projections between V I and MT and V I and V II are illustrated in Fig. 5.5.

Although projections from V I to V II and MT are roughly topological, they do not seem to be strictly point-to-point. The spread of the region of terminations in V II and MT from restricted locations in V I is often clearly greater than what would be expected from the electrophysiological maps. For example, a circumscribed injection site in V I sometimes results in an elongated zone of label extending across much of V II (e.g., ref. 86). Likewise, an injection site in V I often labels a larger proportion of MT than would be expected from the size of the injection site (e.g., refs. 80, 86). Although such an enlarged projection zone in V II could perhaps be explained by the distorted representation of the visual field in V II (21), an alternative explanation is that the striate cortex projec-

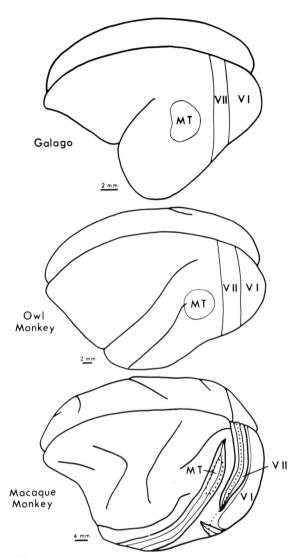

FIG. 5.4. Three basic visual areas in primates. Areas V I, V II and MT have been identified in a range of primates and appear to be basic to the primate visual system. In the prosimian galago, electrophysiological (8; Allman and Kaas, unpublished) and anatomical (86) studies have clearly characterized all three areas. Similarly, these methods have defined these areas in New World owl monkeys (2, 3, 4, 44, Wall and Kaas, unpublished) and other New World monkeys (22, 57, 79, 80, 82, 83, 85, 88, 90, 108, 109). In Old World monkeys, only V I has been easy to identify. V I has clear architectonic boundaries and its visuotopic organization has also been described (25). The organization of V II is incompletely known. The rostral border of V II shown here (dots) is based on our anatomical studies of striate cortex projections in macaque monkeys (102). The rostral border of

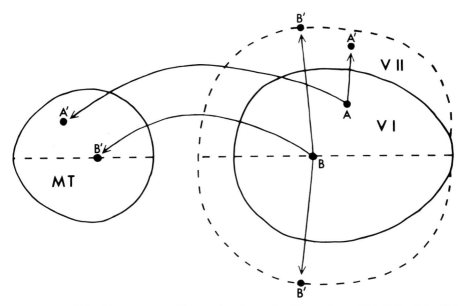

FIG. 5.5. Two types of homotopic projections from V I. Site A in V I projects to a single location in V II and in MT, A′. This type of projection is expected for most parts of V I. However, for positions along all but the most central representation of the horizontal meridian (dashed lines) in V I, projections are to two locations in V II and to one location in MT, as shown by sites B and B′. The reason for this difference is that V I and MT are both simple topographic or first-order representations of the visual hemifield, whereas V II is split along most of the horizontal meridian, so the upper and lower visual quadrants are separated in what is called a second-order representation or transformation (4). Thus, single locations along the horizontal meridian in V I can project to two homotopic locations in V II. Conventions as in Fig. 5.2.

tions diverge in V II. As a further complication, single injection sites in V I have often been shown to result in uneven and discontinuous patches of terminations in V II and MT (57, 102, 109). Such patches of terminations may serve to segregate certain functional inputs in some unknown way. Both the enlarged projection zone from striate cortex and its often discontinuous nature indicate that projection

FIG. 5.4 (*continued*): foveal V II was not determined in this study and remains uncertain. The border of MT (dots) is based on our studies of connections with striate cortex and on myeloarchitecture (99, 102). The location of this striate cortex projection zone is in basic agreement with results of others (23, 28, 49, 76, 94, 95, 113, 115). Because it was necessary to schematically open the lunate sulcus (caudal), inferior occipital sulcus (ventral) and the superior temporal sulcus (rostral) to show the locations of V II and MT in the macaque monkey, the exposed banks of these sulci do not conform to the scales indicated. Conventions as in Fig. 5.2.

patterns exceed those predicted by the excitatory receptive fields of neurons within the mapped visuotopic representation. For this reason, it is often difficult to use patterns of connections to establish the detailed retinotopic organization of areas and to use patterns of retinotopic organization based on connections to subdivide cortex.

Although striate cortex has been shown to project to V II and MT in all the primates that have been studied, projections to additional cortical areas have been inconsistently noted. In part, this may reflect species differences. In addition, these projections may be weak and variable and, therefore, difficult to reveal. Finally, inconsistent evidence may be artifactual. In galagos, V I projections to V II and MT were consistently found, but evidence for slight additional projections to cortex rostral to V II was sometimes noted (86). In New World monkeys, striate cortex projections to only V II and MT have generally been described but Martinez-Millán and Holländer (57) reported that parts of V I representing paracentral and peripheral vision projected to an additional location rostral to V II on the medial wall of the cerebral hemisphere. Strongest evidence of a projection from V I to cortical areas other than V II and MT comes from studies in macaque monkeys. Early lesion and degeneration studies by Cragg and Ainsworth (23) and Zeki (111, 113) described striate cortex projections to MT, V II and a narrow zone of cortex concentrically rostral to V II, which Zeki termed "V III" (111). This striate projection to a "V III" has been more recently supported by Zeki and striate cortex projections to part of even a fourth visual area, "V4(V IV)," have also been described (116).

Since a projection from V I to "V III" and the existence of a concentric type of V III surrounding V II have been postulated as a major difference between Old World and New World monkeys (82), the issue of projections to "V III" is worth considering further. In macaque monkeys, V III has been described as a narrow band of cortex along the rostral border of V II having a mirror image topographical organization to that of V II (96, 111, 117, 119). In our material (102), establishing the validity of "V III" in macaque monkeys from striate cortex projections has proven to be extremely difficult. Projections from single locations in striate cortex not on the horizontal meridian often resulted in two nearby regions of terminations within an adjoining 10 mm wide band of cortex rostral to V I. However, because striate cortex projections, as well as other visual system pathways, have been sometimes shown to be patchy or discontinuous (10, 13, 24, 38, 41, 66, 69, 107–109), such a terminal field may be interpreted as V I projections to patches within V II rather than as projections to both a V II and the very narrow V III. Whether or not the more distant of these foci represents a third visual area, or part of V II, is difficult to determine by such anatomical studies alone. Presently, we are uncertain from our own data and from published reports whether a separate V III projection zone in macaque monkeys is justified.

An alternative position is that V III exists in New World monkeys. If similar proportional widths of V II and V III are proposed for both Old and New World monkeys, then V III, which is about ¼–⅓ the width of V II in Old World monkeys (119), would be only 1 or 2 mm wide in the New World owl monkey (4). We must consider the possibility that such an extremely narrow "V III" has been undetected in both recording experiments (5, 6) and in investigations of area 17 projections in other species of New World (7, 43, 57, 82, 85, 88, 108) and prosimian primates (86, 89). Thus, the presence or absence of a "V III" might or might not be a major difference between Old and New World monkeys.

In Old World monkeys, striate cortex has also been recently described as projecting to an additional cortical area besides V II, "V III" and MT. According to Zeki (116), foveal V I projects to part of a fourth complex of visual areas, V4. The reasons for considering this a projection to part of "V4" rather than to an expansion of V II are given by Zeki (116) and Van Essen and Zeki (96), but none seem so compelling that a major species difference between Old and New World monkeys needs to be accepted. The conservative conclusion for the present seems to be that projections from V I to V II and MT are basic to the organization of the primate visual system, that other projections have been postulated, but not firmly established, and that there may be important species differences in the targets of these additional projections, if they exist.

5. Projections of V II

Although the projections of V I in primates have been studied by numerous investigators, the projections of other cortical visual areas, including V II, have not been extensively investigated. In prosimians, the projections of V II have not been directly studied, but injections of horseradish peroxidase into V I have been shown to retrogradely label neurons in V II, demonstrating a projection to V I (86). Such reciprocal projections back to V I from V II have been clearly demonstrated in both New World (44, 88, 90, 91, 108, 109) and Old World monkeys (99, 102, 109), and therefore seem to be a common feature of the primate visual system.

Studies in New World primates have shown that V II projects to a number of cortical visual areas other than V I. The rostral projections of V II in owl monkeys include a major input to DL and lesser inputs to DM and perhaps MT (44). Similar terminations in locations comparable to these extrastriate areas of the owl monkey have also been described for the squirrel monkey, as well as an additional projection to the frontal eye fields (90). In Old World primates, Zeki (114, 115) has described "V II" projections to what appears to be MT and to areas he terms V3, V3A and to at least two regions within V4, results which at least agree in general with those in New World monkeys, which show that V II projects to a number of other extrastriate visual areas. It is uncertain if any of these visual areas of macaques correspond to the visual areas receiving input from V II in owl monkeys, but our preliminary investigations in macaque monkeys indicate a projection from V II to cortex bordering MT in the expected position of DL (Weller and Kaas, unpublished).

To the extent that the ipsilateral projections of V II to other cortical areas are homotopic, this pattern of projection can be used to address an important anatomical issue. As discussed in detail elsewhere (44, 89), the pattern of V II projections in owl monkeys provides anatomical evidence against the concept of a single "V III" rather than a number of visual areas bordering V II rostrally. As shown in Fig. 5.6, a different pattern of homotopic projections would be expected from V II to DL than from V II to "V III." Thus, the pattern of connections can be used to test the hypothesis developed from electrophysiological mapping studies (5) that cortex between V II and MT is a representation around MT, i.e., DL, rather than one around V II, i.e., V III. It is important to note that the expected projections from some other locations in V II (A in Fig. 5.6) would be the same with either hypothesis, while the expected projections from other locations (B and more so in C) would differ. The actual results in owl monkeys were consistent with the concept of DL rather than V III (44). Studies of V II projections in the squirrel monkey (89, 109) have not resolved this issue, as pointed out by Tigges, et al. (89), because the projections of only the part of V II roughly comparable to location A in Fig. 5.6 have been investigated. In macaque monkeys, projections of "V II" have only been studied using large cortical lesions that are defined by location only as involving "V II" (23, 49, 114) and the cortical lesions have typically involved cortex buried in fissures found only on the lateral surface of the brain, again roughly comparable to location A of Fig. 5.6.

6. Projections of MT

MT has been shown to be reciprocally connected with V I in a range of primates. Most information has been obtained from injections of horseradish peroxidase into striate cortex. Injections of horserad-

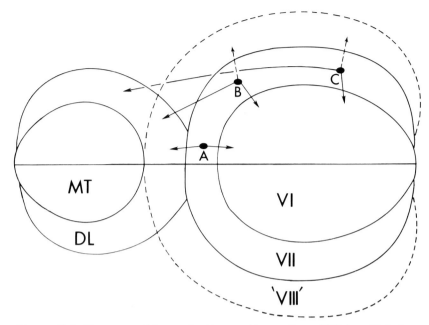

Fig. 5.6 Two possible projection patterns from V II to adjoining cortex in primates. The expected projections from the portion of VII representing central vision (A) to homotopic locations in V III or in the dorsolateral visual area (DL) would be similar and would be consistent with either hypothesis. Projections from parts of V II representing paracentral (B) and peripheral (C) vision to homotopic locations in V III or DL would be different. In owl monkeys, the projecton pattern supported the concept of DL rather than V III. Modified from Kaas and Lin (44).

ish peroxidase in V I in prosimian galagos (86), New World marmosets (79, 80) and squirrel monkeys (109) and Old World macaque monkeys (99, 102, 109) have demonstrated projections from MT. These projections have been confirmed by the use of anterograde tracers or lesions in MT in marmosets (80, 83), owl monkeys (Wall and Kaas, unpublished) and galagos (98). From these experiments, we can postulate that a projection to striate cortex is a general feature of MT in primates.

The other cortical projections of MT are less well established. Spatz and Tigges (83) found evidence for a number of additional projection zones of MT in the marmoset, including locations that appear comparable to the owl monkey extrastriate visual areas DL, DM, M and PP, and the frontal eye fields. Our studies in owl monkeys (Wall and Kaas, unpublished) indicate that MT projects to DL, DM, M and possibly other visual areas. In galagos, MT also projects to several extrastriate cortical sites (98). Finally, projections of temporal cortex in the region of MT in macaque monkeys (49, 58, 67) are consistent with the concept that MT sends efferents to a number of other cortical areas.

7. Projections of Other Visual Areas in the Owl Monkey

In addition to V I, V II and MT, the connections of several other regions of visual cortex have been studied in owl monkeys. The efferent pathways of DM, M and PP have been studied with autoradiographic tracing techniques in owl monkeys. A major output of DM is to posterior parietal cortex, PP, just anterior to DM. Other projections of DM are to MT, DL, M and an undefined area on the medial wall of the cerebral hemisphere (97). PP sends efferents to DM, M, MT, DL, the frontal eye fields, other subdivisions of parietal cortex and other locations in visually responsive cortex (46). Area M projects to V II, DM, MT, DL, DI, cortex rostral to M on the medial wall and other visually responsive cortex (31). Although the projections of the remaining extrastriate visual areas of the owl monkey have not been completely studied, preliminary results show that DL projects to numerous extrastriate visual areas, including sites in the inferior temporal lobe (100). In addition, injections of horseradish peroxidase into MT have indicated DL projections to MT (Wall and Kaas, unpublished). The known ipsilateral cortical connections of cortical visual areas in owl monkeys are summarized in Fig. 5.7.

8. Corpus Callosum Projections

The contralateral cortical projections of the numerous visual areas are also complex. There is evidence for the three types of callosal connections shown schematically in Fig. 5.8: homotopic, homoregional and heteroregional (43). Homotopic connections are the ones usually considered in descriptions of callosal projections. Homotopic projections connect matched points along the representation of the vertical meridian in pairs of visual areas, so that neurons in the two hemispheres with similar receptive fields can interrelate. These connections bridge the hemifield representations in the two hemispheres, are the equivalent of shorter connections within hemifield representations and probably relate to the border regions of visual representations, wherever the border corresponds to the vertical meridian. Homotopic callosal connections for visual areas have been clearly demonstrated for the region of the V I border

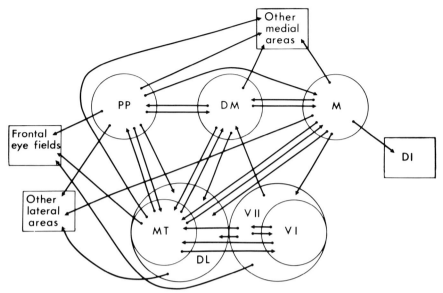

FIG. 5.7. Some connections of visual cortex in owl monkeys. Conclusions are based on results in owl monkeys (31, 44, 46, 97 and Wall and Kaas, unpublished) and are supported by results in other New World monkeys (57, 80, 82, 83, 85, 88, 91, 108, 109). Since the organization of visual cortex is incompletely understood, several subdivisions of cortex are probably grouped in the lateral areas, medial areas and frontal eye fields. Conventions as in Figs. 5.2 and 5.3.

with V II (44, 61, 85, 90, 96, 103, 105, 112), where they are obvious, since V I and V II both appear to lack the other two types of callosal connections. In owl monkeys, corpus callosum connections are also concentrated around the border of MT with DL (62). Since this border represents the vertical meridian, these connections are most probably homotopic.

Homoregional corpus callosum projections are from the central regions of a representation to the central regions of the corresponding representation in the other hemisphere. Areas V I and V II do not have this second kind of projection, and limited evidence from injections in Area M in the owl monkey suggests that this area does not also (31). However, injections or lesions centered in MT (83; Wall and Kaas, unpublished), PP (46), DM (97) or DL (100) revealed terminations centered in the contralateral MT, PP, DM or DL, respectively. Since the connections are between the representations of nonoverlapping parts of the right and left hemifields, the connections cannot be homotopic, so we use the term homoregional, because they connect the same visual area of each hemisphere. Presumably, homoregional connections allow stimuli in one hemifield

A. Homotopic

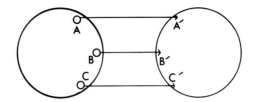

B. Homoregional

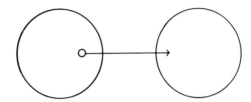

C. Heteroregional

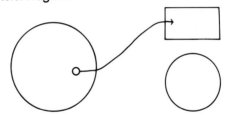

FIG. 5.8. Three types of interhemispheric connections. A, homo-topic pathways connect identical or nearly identical visual field locations along the representations of the zero vertical meridian (line of decussa-tion) in the same visual areas of the right and left cerebral hemispheres. B, homoregional pathways connect nonidentical, but perhaps mirror, loca-tions in the two hemifield representations in the same visual areas of the two hemispheres. C, heteroregional pathways connect one visual area of one hemisphere with different visual areas in the other hemisphere. Al-though V I, V II and area M may have only homotopic callosal connections (31, 44, 86, 90, 96, 103, 105, 112), other visual areas, such as MT (83) and DM (97), appear to have all three types.

to influence the responses of neurons activated by stimuli in mirror symmetrical portions of the other hemifield, but the electrophysio-logical consequences of homoregional corpus callosum connec-tions have not yet been demonstrated (except for inferotemporal cortex, where such connections produce large bilateral receptive fields; 70).

A third type of corpus callosum connection is heteroregional. Just as a visual area may have several ipsilateral connections with other cortical visual areas, some visual areas connect contralaterally not only with their counterparts but with other visual areas as well. So far, it has been shown that DM projects heteroregionally to MT and PP of the opposite hemisphere (97), MT projects heteroregionally to DL (Wall and Kaas, unpublished; also see ref. 83) and DL projects heteroregionally to the inferior temporal lobe (100). The functional role of such connections is not known, but heteroregional callosal connections certainly add to the multitude of influences impinging on some cortical visual areas.

Perhaps some of the extrastriate visual areas in macaque monkeys also have homoregional or heteroregional callosal connections, since, following corpus callosum section, much of extrastriate cortex rostral to V II contains large areas of terminal degeneration (96, 112). Sectioning the corpus callosum does not allow a distinction between types of callosal connections, however, but only indicates the overall presence of any type of connection.

9. Connections of Subdivisions of Visual Cortex with Subcortical Structures

Each cortical visual area is connected with numerous subcortical structures in a complex way. Individual visual areas have many subcortical connections in common and yet each visual area appears to have a unique pattern of subcortical connections. As for the cortical connections between visual areas, more is known about the subcortical connections of V I in a range of primates than for any other visual area. Extrastriate visual areas have not been subject to as many anatomical studies, so less is known about their subcortical connections.

Table 5.1 lists the subcortical targets of six visual areas in the owl monkey (modified from 32) and gives an impression of the complexity of the subcortical projection pattern that presumably would exist in any primate. Some of these connections are discussed further below.

9.1. Cortical Projections to the Lateral Geniculate Nucleus and Superior Colliculus

It appears that all significant visual information to cortex is relayed through the lateral geniculate nucleus and the superior colliculus. These targets of direct projections from the retina also receive major projections from visual areas of cortex and, therefore, the activity evoked by retinal input is subject to cortical control and mod-

Table 5.1
Summary of Projections (×) Found from Six Visual
Cortical Areas in the Owl Monkey to Subcortical Structures.
Modified from Graham et al. (32).

Brain structure	VI	VII	MT	DM	M	PP
Caudate and Putamen			×	×	×	×
Claustrum				×	×	×
Reticular nucleus	×	×	×	×	×	×
Dorsal lateral geniculate nucleus	×	×	×			
Pregeniculate nucleus				×	×	×
Inferior pulvinar, central	×	×	×	×	×	×
medial	×	×	×	×	×	×
posterior						
Superior pulvinar, lateral	×	×	×	×	×	×
central			×	×	×	×
Lateral posterior nucleus				×	×	×
Intralaminar nuclei					×	×
Zona incerta						×
Posterior pretectal nucleus and/or nucleus of the optic tract	×	×	×	×	×	×
Anterior pretectal nucleus					×	×
Superior colliculus	×	×	×	×	×	×
Pontine nuclei			×	×	×	×

ulation. The lateral geniculate nucleus is the major source of visual input to cortex and its cortical target, striate cortex, projects back strongly to the lateral geniculate nucleus, with some emphasis on interlaminar zone terminations (32, 38, 39, 40, 50, 81, 86). More recently, it has become apparent that V II and MT also project to the lateral geniculate nucleus. Injections of ^{3}H-proline in V II and MT of the owl monkey resulted in concentrations of label in the magnocellular layers, the superficial layers and the adjoining interlaminar zones of the nucleus (32, 50). Injections into "area 18" of macaque monkeys also revealed projections to the lateral geniculate nucleus (38, 106) and injections of horseradish peroxidase into the geniculate retrogradely labeled cells in both "area 18" and the region of MT (38). Results from both types of investigations indicate that the input from area 18 to the lateral geniculate nucleus is relatively sparse.

The superior colliculus relays visual information to visual cortical areas through its projections to subdivisions of the pulvinar complex. It is not known whether all visual cortical areas project to the superior colliculus, but in a recent investigation of the subcortical projections of six visual areas in owl monkeys, all were found to

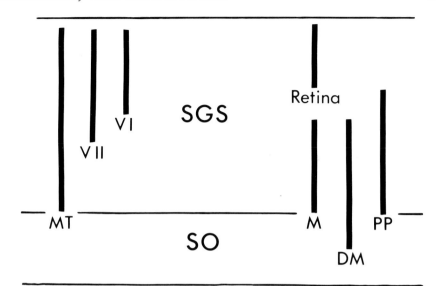

FIG. 5.9. Tectal inputs from visual areas of cortex in the owl monkey vary in depth. A section of the superficial superior colliculus (stratum griseum superifciale-SGS; stratum opticum-SO) perpendicular to its surface with vertical bars indicating the extents and depths of cortical projections; retinal projections are shown for comparison. Modified from Graham et al. (32). Conventions after Fig. 5.2 and 5.3.

have substantial tectal projections (32). Projections from V I (55, 65, 85, 86, 89, 104) and extrastriate visual areas (19, 53, 84) to the superior colliculus have also been found in other primates. These tectal projections from different cortical areas in the owl monkey vary in depth of termination in the superior colliculus, as shown in Fig. 5.9. It is interesting that the visual areas receiving the most direct visual information relayed through the lateral geniculate nucleus (V I, V II and MT) project the most superficially in the superior colliculus and have extensive overlap with the retinal input. The V I terminations are the most superficial. Visual areas less directly related to the visual input carried by the geniculostriate system (DM, PP and M) project to deeper locations in the superior colliculus. Since the deeper tectal layers function in directing eye movements towards visual stimuli (60, 72, 75, 78, 110), the deeper terminations from cortex may be more directly involved in controlling eye movements.

9.2. *Projections to the Pulvinar Complex*

Patterns of connections with cortex, as well as other anatomical and some electrophysiological data, suggest that the inferior and superior divisions of pulvinar can be further subdivided into a number of nuclei, each with its own pattern of visual connections. Thus, the inferior pulvinar of the owl monkey has been divided into three nuclei, medial, lateral and central, based on differences in patterns of connections and architectonic appearance (51). One of these subdivisions has also been electrophysiologically investigated and shown to contain a topographic representation of visual space (9, see 29 for results in cebus monkeys), whereas anatomical studies suggest that the other subdivisions also contain such orderly representations. Likewise, the superior pulvinar of owl monkeys has also been divided into at least two divisions by its patterns of connections (32). As noted in Table 5.1, each division of the inferior pulvinar and the superior pulvinar with visual input receives projections from several cortical visual areas (32, 51). The major output of these nuclei is back to subdivisions of visual cortex (see 51 for review). Thus, the nuclei of the pulvinar complex serve to interrelate subdivisions of visual cortex.

Similar conclusions can be reached by studying cortical–pulvinar relations in other primates. The projections of V I (39, 65, 81), V II (107) and MT (84) in other New World monkeys, the projections of V I in galagos (86) and projections of V I and other visual cortical regions in Old World monkeys (10, 18, 32, 55, 65, 69) also suggest that visual cortical areas project to nuclei within the pulvinar complex in a pattern at least roughly similar to that in owl monkeys.

9.3. Projections to the Pregeniculate Nucleus and the Reticular Nucleus of the Thalamus

All known subdivisions of visual cortex in primates project to the caudal portion of the reticular nucleus (17, 19, 32, 39, 65, 81, 84, 86) and at least some project to the pregeniculate nucleus (17, 19, 32, 39, 65, 81, 84, 85). The reticular nucleus in turn projects to the lateral geniculate nucleus (42), so that visual cortex also has this indirect route for influencing the activity of the lateral geniculate nucleus. Although widely separate regions of cortex are known to have separate termination zones in the reticular nucleus, it is not yet certain if adjacent visual areas may have separate termination zones in the reticular nucleus or overlap.

9.4. Projections to the Basal Ganglia

Various subdivisions of visual cortex are sometimes reported to send a sparse projection to locations within the basal ganglia, including the caudate, putamen and claustrum (10, 32, 46, 64, 84,

86, 97). The functions of the connections of the cortical visual areas with the basal ganglia are not understood, but it seems probable that they are important in the direction and control of complex motor activity. It is perhaps significant that although striate cortex projections to the basal ganglia are sparse if present (86), projections from extrastriate visual areas are marked (see 32 for review).

9.5. Projections to the Pons

In primates, V I does not appear to have direct connections with nuclei in the pons (e.g., 32, 86). Most studies reporting such a projection have tended to involve more than just V I (14, 17, 19, 85). Extrastriate visual cortical areas, in contrast, do have obvious pontine projections. In owl monkeys, DM, M, MT and PP (32, 46, 64) have been shown to project to regions of the pons. A projection to the pons was found to be the most pronounced of the subcortical projections from MT in squirrel monkeys (84). The projections of visual areas to the pons are similar in Old World monkeys to those reported for the New World monkeys to the extent that they come from wide regions of extrastriate cortex (14, 19, 53), and thus probably involve a number of visual areas. Because regions of the pons receiving visual cortical input in turn project to the cerebellum (30), the visual corticopontine connections may be important in the regulation of eye, head and body movements relevant to vision.

10. Reciprocal and Nonreciprocal Connections

Most connections of cortical visual areas with each other and with thalamic nuclei are reciprocal and homotopic, relating similar portions of the visual field in the separate representations (see Fig. 5.7). Thus, V I projects to V II and MT, and both these visual areas project back to V I (see figure legend for these and following references). Likewise, V I receives input from the lateral geniculate nucleus, which itself is the target of a major returning pathway from V I. Another example of reciprocal corticothalamic relations is the strong interconnection between MT and the medial nucleus of the inferior pulvinar complex in owl monkeys.

Though reciprocal connections between visual structures are common, they are not universal. MT and V II project to the lateral geniculate nucleus, for example, whereas the lateral geniculate nucleus does not project to MT or V II. Of course, many brain stem targets of cortical areas, such as the reticular nucleus, the pregeniculate nucleus, the superior colliculus and the pons, are nonreciprocal in that these structures have no cortical projections. Instances of nonreciprocal pathways can also be found in the connections of cortical visual areas. Area M projects to V II, but V II apparently does not project to M. Yet, visual structures that are not directly related via reciprocal connections can still be related via projections through another common structure. Projections involving the lateral or superior pulvinar could play this role (10, 13, 69). However, our present understanding of the functional roles of connections is so lacking that a general explanation of why some connections are reciprocal and others are not is not yet possible.

11. Laminar Patterns of Connections

Connections of visual cortical areas originate and terminate in different layers, as shown in Fig. 5.10, and such laminar patterns of connections have functional implications. Based on anatomical studies in the owl monkey, there appear to be at least four variations in the patterns of corticocortical terminations within the subdivisions of visual cortex. The most common pattern is a concentration of terminations in layers III and IV, as seen in the projections of V I to V II, the projections of V II to DL, DM and MT, PP inputs to MT, DL, M and the frontal eye fields, DM inputs to MT, DL and PP, and MT inputs to DL, DM, M, PP and the frontal eye fields (see figure legend for these and following references). These afferents presumably synapse with the numerous small intrinsically projecting granule cells of layers III and IV. Since these neurons also receive incoming axons from the thalamus, terminations in layers III and IV presumably are to early stages of the local circuit cortical processing. It is apparently important to have multiple influences at these early stages. A second type of corticocortical projection pattern is shown by V II projections to V I and in PP projections to DM that terminate largely in cortical layers V and I. This pattern of terminations is positioned to affect the apical dendrites in layer I and the basal dendrites in layer V of pyramidal cells of layer V. Since these pyramidal cells are efferent neurons, this second less-frequent pattern of terminations could be a system for directly modulating the final output of cortical areas. A third pattern of corticocortical terminations in primates has been seen only for the projections of area M (31). Area M projections to other areas such as V II and MT appear to be concentrated in layers VI and to a lesser extent I. Layer VI is commonly the source of reciprocal efferents to the thalamic nucleus providing the major thalamic input to the cortical area. An interesting exception is that layer VI cells of MT project to V I (55, 86, 99)

Afferents

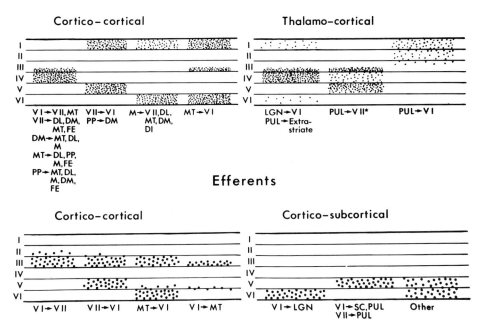

Cortico-cortical   Thalamo-cortical

Efferents

Cortico-cortical   Cortico-subcortical

FIG. 5.10. Laminar patterns of afferent terminations and efferent neurons for visual areas of cortex. Small dots indicate projections into cortex; larger dots represent efferent cells of origin. The details of the corticocortical afferents are from studies in the owl monkey (31, 44, 46, 97, Wall and Kaas, unpublished studies) and projections of V I (57, 68, 80, 85, 86, 99, 102, 108, 109), V II (88, 89, 91, 102, 108, 109) and MT (80, 83, 99, 102) in other primates. Thalamocortical afferent patterns are known for the geniculostriate pathway (38, 41, 69, 91), the pulvinoextrastriate pathway (13, 66, 92) and the pulvinostriate pathway (12, 13, 20, 66, 69). The * denotes a different pattern seen for pulvinar projections to V II in the squirrel monkey (24). The laminar locations of the cells of origin of efferent corticocortical projections for V I to V II are taken from refs. 57, 99, 108, 109, for V II to V I from refs. 86, 99, 109, for MT to V I from refs. 55, 80, 86, 99, 109 and for V I to MT from 79, 80 and Wall and Kaas, unpublished. The laminar patterns of the corticosubcortical efferents for V I to LGN are from refs. 37, 54, 67, V I to SC and pulvinar from ref. 55, V II to pulvinar from ref. 68 and remaining projections to subcortical structures from ref. 93. Frontal eye fields (FE), pulvinar complex (PUL), lateral geniculate nucleus (LGN), superior colliculus (SC), projections to either various or undefined subcortical targets (Other). Other conventions as in Figs. 5.2 and 5.3.

but, again, the layer VI cells are projecting in a reciprocal manner back to an earlier station in a serial processing sequence of visual input to MT. The area M projections into layer VI, therefore, complement the layer V terminations and modulate the layer VI output of cortical areas. Finally, another variant of laminar terminations is seen in MT projections to V I, which are found in layers I, deeper III and VI (83; Wall and Kaas, unpublished).

The thalamic projections to visual cortical areas also have differences in the laminar pattern of their terminations. As previously mentioned, inputs from the thalamic nucleus providing a major projection to a cortical area terminate in layers III and IV. Thus, the lateral geniculate nucleus projects to layers III and IV of striate cortex (see 38 for review) and the pulvinar nuclei project to layers III and IV of extrastriate cortex (see 69 for review). A slightly different result has been described for pulvinar projections to area 18 in the squirrel monkey, from which projections are primarily to layers III and V (24). A second major pattern of terminations is found when a thalamic nucleus sends a sparse projection to a cortical area not considered its primary target. These terminations appear in layers above those of the major thalamic afferents, or in layers I, II, and sometimes the upper part of III. Pulvinar nuclei that send a sparse projection to V I terminate in these supragranular layers (66, 69) and "nonspecific" thalamic structures, such as the intralaminar nuclei, have been traditionally described as terminating in layer I (20, 48, 54).

Neurons that project efferently from visual cortex are found in layers III, V and V I (see Fig. 5.10), although in specific areas the details of the locations of these efferent cells vary. For all subdivisions of visual cortex in primates for which information is available, cells projecting to other cortical areas are found in layer III and to a lesser extent in layer V and sometimes layer V I. Cells projecting subcortically are found in layers V and VI. For example, pyramidal cells in layers III and V of area 18 or V II are retrogradely labeled following injection of HRP into V I (86, 99, 102, 109), while cells in layers VI and V of V I are labeled following such injections into the lateral geniculate nucleus and the pulvinar, respectively (38, 55).

12. Conclusions

Connections between cortical and subcortical components of the primate visual system are remarkably numerous and complex. An area of visual cortex receives major afferent input from a specific thalamic nucleus, additional thalamic projections and a myriad of cortical afferents. In addition, an area projects to numerous brainstem structures, not all of which are traditional visual structures, and projects to numerous other visual cortical areas. Connections between cortical areas are laminarly organized and commonly, but not always, reciprocal. Not only does visual cortex project ipsilaterally to cortical and subcortical areas, but different subdivisions have different types of callosal projections to contralateral visual areas of cortex. At least most visual structures contain a retinotopic representation of visual space, and connections between visual areas and subcortical nuclei are topographically remarkably specific. The number of visual areas and their connections have not yet been fully determined in any primate and future work will undoubtedly add to the known complexity of cortical organization.

The present understanding of cortical and subcortical connections in the visual systems of primates allows several conclusions. First, the complicated pathways between visual structures argue against the traditional strictly serial view of visual processing and emphasize, rather, the constant multiple interactions and modulations that must occur between visual structures. Second, at least some visual areas and connections appear to be common to all primates. Three visual areas, V I, V II and MT, have been found in all the primates studied and future studies may add to this number. Not only are these areas common to primates, but many connections involving them are also shared, such as the reciprocal pathway between V I and MT. In addition, projections from V II and MT to other regions of cortex indicate that multiple extrastriate visual areas are probably present in all primates. Subdivisions of the pulvinar complex also seem to exist in all primates, although many of the details of such subdivisions and their connections with cortex are presently unclear. A third conclusion suggested by a description of the pathways between visual structures in primates is that areas and nuclei are linked into functionally distinct subsystems. We have mentioned parallel pathways from retina to cortex related to the X, Y and W cell systems. Not only are there separate channels carrying visual information to cortex, but interconnections between visual areas and subcortical structures may maintain this functional segregation to some degree. The Y cell system includes retinal projections to both the magnocellular layers of the LGN and to the superior colliculus. Tectal cells project back to the magnocellular region of the lateral geniculate nucleus (36) and cortical visual areas V II and MT also project to the magnocellular region. All these connections could be largely related to Y cell activity. Other subsystems of visual processing may exist besides those related to the type of afferent information transmitted. Based on a study of subcortical projections of various areas of visual cortex in owl monkeys (32), we suggested that V I, V II and MT were functionally interrelated, based on their shared connections, as were areas DM, M and PP. An obvious distinction between these two groupings of cortical areas is their relation to direct visual afferent input. Striate cortex is the only visual area to receive a direct projection from the lateral geniculate nucleus in primates and it relays that visual input to V II and MT. Areas DM, M and PP, in contrast, receive cortical input from extrastriate visual areas and have subcortical targets in addition to those to which V I, V II and MT project.

References

1. ALLMAN, J. M. Cortical visual areas in the owl monkey: topographic organization and functional correlates. This volume, chapter 15.
2. ALLMAN, J. M., AND KAAS, J. H. A representation of the visual field in the caudal third of the middle temporal gyrus of the owl monkey (*Aotus trivirgatus*). *Brain Res.*, 31: 85–105, 1971a.
3. ALLMAN, J. M., AND KAAS, J. H. Representation of the visual field in striate and adjoining cortex of the owl monkey (*Aotus trivirgatus*). *Brain Res.*, 35: 89–106, 1971b.
4. ALLMAN, J. M., AND KAAS, J. H. The organization of the second visual area (V II) in the owl monkey: A second order transformation of the visual hemifield. *Brain Res.*, 81: 247–265, 1974a.
5. ALLMAN, J. M., AND KAAS, J. H. A crescent-shaped cortical visual area surrounding the middle temporal area (MT) in the owl monkey (*Aotus trivirgatus*). *Brain Res.*, 81: 199–213, 1974b.
6. ALLMAN, J. M., AND KAAS, J. H. The dorsomedial cortical visual area: A third tier area in the occipital lobe of the owl monkey (*Aotus trivirgatus*). *Brain Res.*, 100: 473–487, 1975.
7. ALLMAN, J. M., AND KAAS, J. H. Representation of the visual field on the medial wall of occipital-parietal cortex in the owl monkey. *Science*, 191: 572–575, 1976.
8. ALLMAN, J. M., AND KAAS, J. H., AND LANE, R. H. The middle temporal visual area (MT) in the bushbaby, *Galago senegalensis*. *Brain Res.*, 57: 197–202, 1973.
9. ALLMAN, J. M., AND KAAS, J. H., AND MIEZIN, F. M. A representation of the visual field in the inferior nucleus of the pulvinar in the owl monkey (*Aotus trivirgatus*). *Brain Res.*, 40: 291–302, 1972.
10. BENEVENTO, L. A., AND DAVIS, B. Topographical projections of the prestriate cortex to the pulvinar nuclei in the macaque monkey: an autoradiographic study. *Exptl. Brain Res.*, 30: 405–424, 1977.
11. BENEVENTO, L. A., AND FALLON, J. The ascending projections of the superior colliculus in the rhesus monkey (*Macaca mulatta*). *J. Comp. Neurol.*, 160: 339–362, 1975.
12. BENEVENTO, L. A., AND REZAK, M. Extrageniculate projections to layers VI and I of striate cortex (area 17) in the rhesus monkey (*Macaca mulatta*). *Brain Res.*, 96: 51–55, 1975.
13. BENEVENTO, L. A., AND REZAK, M. The cortical projections of the inferior pulvinar and adjacent lateral pulvinar in the rhesus monkey

(*Macaca mulatta*): An autoradiographic study. *Brain Res.*, 108: 1–24, 1976.

14. BRODAL, P. The corticopontine projection in the rhesus monkey. *Brain*, 101: 251–283, 1978.

15. BRODMANN, K. Beitrage zur histologischen Lokalisation der Grosshirnrinde. *J. Psych. Neurol., Leipzig*, 4: 177–226, 1905.

16. BUNT, A., HENDRICKSON, A., LUND, J., LUND, R., AND FUCHS, A. Monkey retinal ganglion cells: morphometric analysis and tracing of axonal projections, with a consideration of the peroxidase technique, *J. Comp. Neurol.*, 164: 265–286, 1975.

17. CAMPOS-ORTEGA, J. A. Descending subcortical projections from the occipital lobe of *Galago crassicaudatus. Exptl. Neurol.*, 21: 440–454, 1968.

18. CAMPOS-ORTEGA, J. A., AND HAYHOW, W. R. On the organization of the visual cortical projection to the pulvinar in *Macaca mulatta. Brain, Behav. Evol.*, 6: 394–443, 1972.

19. CAMPOS-ORTEGA, J. A., HAYHOW, W. R., AND CULVER, P. F. DeV. The descending projections from the cortical visual fields of *Macaca mulatta* with particular reference to the question of a cortico–lateral geniculate pathway. *Brain, Behav. Evol.*, 3: 368–414, 1970.

20. CAREY, R. G., FITZPATRICK, P., AND DIAMOND, I. T. Layer I of striate cortex of *Tupaia glis* and *Galago senegalensis:* Projections from the thalamus and claustrum revealed by retrograde transport of horseradish peroxidase. *J. Comp. Neurol.*, 186: 393–438, 1979.

21. COLONNIER, M., AND SAS, E. An anterograde degeneration study of the tangential spread of axons in cortical areas 17 and 18 of the squirrel monkey (*Saimiri sciureus*). *J. Comp. Neurol.*, 179: 245–262, 1978.

22. COWEY, A. Projection of the retina on the striate cortex and prestriate cortex in the squirrel monkey, *Saimiri sciureus, J. Neurophysiol.*, 27: 266–393, 1964.

23. CRAGG, B. G., AND AINSWORTH, A. The topography of the afferent projections in the circumstriate visual cortex of the monkey studied by the Nauta method. *Vision Res.*, 9: 733–747, 1969.

24. CURCIO, C. A., AND HARTING, J. K. Organization of pulvinar afferents to area 18 in the squirrel monkey: evidence for stripes. *Brain Res.*, 143: 155–161, 1978.

25. DANIEL, P. M., AND WHITTERIDGE, D. The representation of the visual field on the cerebral cortex in monkeys. *J. Physiol., London*, 159–221, 1961.

26. DIAMOND, I. T., AND HALL, W. C. Evolution of neocortex. *Science*, 164: 251–262, 1969.

27. DREHER, B., FUKADA, Y., AND RODIECK, R. Identification, classification and anatomical segregation of cells with X-like properties in the lateral geniculate nucleus of Old World primates. *J. Physiol., London*, 258: 433–452, 1976.

28. GATTASS, R., AND GROSS, C. G. A visuotopically organized area in

the posterior superior temporal sulcus of the macaque. *ARVO Abstracts,* 184, 1979.

29. GATTASS, R., OSWALDO-CRUZ, E. AND SOUSA, A. P. B. Visuotopic organization of the cebus pulvinar: A double representation of the contralateral hemifield. *Brain Res.,* 152: 1–16, 1978.

30. GLICKSTEIN, M., STEIN, J. AND KING, R. A. Visual input to the pontine nuclei. *Science,* 173: 1110-1111, 1972.

31. GRAHAM, J., WALL, J., AND KAAS, J. H. Cortical projections of the medial visual area in the owl monkey. *Neurosci. Letters,* 15: 109–114, 1979.

32. GRAHAM, J., LIN, C.-S., AND KAAS, J. H. Subcortical projections of six visual cortical areas in the owl monkey, *Aotus trivirgatus. J. Comp. Neurol.,* 187: 557–580, 1979.

33. GROSS, J. C., BENDER, D. B., AND ROCHA-MIRANDA, D. E. Inferotemporal cortex: a single-unit analysis. In: F. O. SCHMITT AND F. G. WORDEN. *The Neurosciences: Third Study Program,* edited by Cambridge, Mass., MIT Press, 1973, 229–238.

34. GROSS, C. G., BRUCE, C. J., DESIMONE, R., FLEMING, J, AND GATTASS, R. Visual areas of the temporal lobe. This volume, chapter 16.

35. HARTING, J. K., HALL, W. C., AND DIAMOND, I. T. Evolution of the pulvinar. *Brain, Behav. Evol.* 6: 424–452, 1972.

36. HARTING, J. K., CASAGRANDE, V. A., AND WEBER, J. T. The projection of the primate superior colliculus upon the dorsal lateral geniculate nucleus: autoradiographic demonstration of interlaminar distribution of tectogeniculate axons. *Brain Res.,* 150: 593–599, 1978.

37. HASSLER, R. Comparative anatomy of the central visual systems in day- and night-active primates. In: Evolution of the Forebrain, edited by R. HASSLER AND H. STEPHAN. New York: Plenum Press, 1967, pp. 419–434.

38. HENDRICKSON, A. E., WILSON, J. R., AND OGREN, M. P. The neuroanatomical organization of pathways between the dorsal lateral geniculate nucleus and visual cortex in Old World and New World primates. *J. Comp. Neurol.,* 182: 123–136, 1978.

39. HOLLÄNDER, H. Projections from the striate cortex to the diencephalon in the squirrel monkey (*Saimiri sciureus*). A light microscopic radioautographic study following intracortical injection of ^{3}H leucine. *J. Comp. Neurol.,* 155: 425–440, 1974.

40. HOLLÄNDER, H., AND MARTINEZ-MILLÁN, M. Autoradiographic evidence for a topographically organized projection from the striate cortex to the lateral geniculate nucleus in the rhesus monkey (*Macaca mulatta*). *Brain Res.,* 100: 407–411, 1975.

41. HUBEL, D. H., AND WIESEL, T. N. Laminar and columnar distribution of geniculocortical fibers in the macaque monkey. *J. Comp. Neurol.,* 146: 421–450, 1972.

42. JONES, E. G. Some aspects of the organization of the thalamic reticular complex. *J. Comp. Neurol.,* 162: 285–308, 1975.

43. KAAS, J. H. The organization of visual cortex in primates. In: *Sensory Systems of Primates*, edited by C. R. NOBACK. New York: Plenum Press, 1978, pp. 151–179.

44. KAAS, J., AND LIN, C.-S. Cortical projections of area 18 in owl monkeys. *Vision Res.*, 17: 739–741, 1977.

45. KAAS, J. H., GUILLERY, R. W., AND ALLMAN, J. M. Some principles of organization in the dorsal lateral geniculate nucleus. *Brain Behav. Evol.*, 6: 253–299, 1972.

46. KAAS, J. H., LIN, C.-S., AND WAGOR, E. Cortical projections of posterior parietal cortex in owl monkeys. *J. Comp. Neurol.*, 171: 387–408, 1977.

47. KAAS, J. H., HUERTA, M. F., WEBER. J.T., AND HARTING, J. K. Patterns of retinal terminations and laminar organization of the lateral geniculate nucleus of primates. *J. Comp. Neurol.*, 182: 517–554, 1978.

48. KILLACKEY, H. P. AND EBNER, F. F. Two different types of thalamocortical projections to a single cortical area in mammals. *Brain, Behav. Evol.*, 6: 141–169, 1972.

49. KUYPERS, H. G., SZWARCBART, M. K., MISHKIN, M., AND ROSVOLD, H. E. Occipitotemporal corticocortical connections in the rhesus monkey. *Exptl. Neurol.*, 11: 245–262, 1965.

50. LIN, C.-S., AND KAAS, J. H. Projections from cortical visual areas 17, 18, and MT onto the dorsal lateral geniculate nucleus in owl monkeys. *J. Comp. Neurol.*, 173: 457–473, 1977.

51. LIN, C.-S., AND KAAS, J. H. The inferior pulvinar complex in owl monkeys: architectonic subdivisions and patterns of input from the superior colliculus and subdivisions of visual cortex. *J. Comp. Neurol.*, 187: 655–678, 1979.

52. LIN, C.-S., WAGOR, E., AND KAAS, J. H. Projections from the pulvinar to the middle temporal visual area (MT) in the owl monkey, *Aotus trivirgatus*. *Brain Res.*, 76: 145–149, 1974.

53. LOCKE, S., WRIGHT, S., JR., AND HILSZ, J. Projection of medial occipital cortex of macaque to posterior thalamus. *Brain*, 97: 65–68, 1974.

54. LORENTE DE Nó. Architectonics and the structure of the cerebral cortex. In: *Physiology of the Nervous System*, Edited by J. F. FULTON. New York: Oxford University Press, 291–330, 1938.

55. LUND, J. S., LUND, R. D., HENDRICKSON, A. E., BUNT, A. H., AND FUCHS, A. F. The origin of efferent pathways from the primary visual cortex, area 17, of the macaque monkey as shown by retrograde transport of horseradish peroxidase. *J. Comp. Neurol.*, 164: 287–304.

56. MAROCCO, R. T., AND LI, R. H. Monkey superior colliculus: properties of single cells and their afferent inputs. *J. Neurophysiol.*, 40: 844–860, 1977.

57. MARTINEZ-MILLÁN, M., AND HOLLÄNDER, H. Cortico-cortical projections from striate cortex of the squirrel monkey (*Saimiri*

sciureus). A radioautographic study. *Brain Res.*, 83: 405–417, 1975.

58. McLoon, S. D., Santos-Anderson, R., and Benevento, L. A. Some projections of the posterior bank and floor of the superior temporal sulcus in the macaque monkey. *Neurosci. Abst.*, 1: 64, 1975.

59. Mishkin, M. Visual mechanisms beyond the striate cortex. In: Russell, R. W. *Frontiers in Physiological Psychology*, New York: Academic Press, 1966, pp. 93–119.

60. Mohler, C. W., and Wurtz, R. H. Organization of monkey superior colliculus: Intermediate layer cells discharging before eye movements. *J. Neurophysiol.*, 39: 722–744, 1976.

61. Myers, R. E. Commissural connections between occipital lobes of the monkey. *J. Comp. Neurol.*, 118: 1–16, 1962.

62. Newsome, W. T., and Allman, J. M. Interhemispheric connections of visual cortex in the owl monkey, *Aotus trivigatus*, and the bushbaby, *Galago sengalensis. J. Comp. Neurol.*, 194: 209–234, 1980.

63. Norden, J. J., and Kaas, J. H. The identification of relay neurons in the dorsal lateral geniculate nucleus of monkeys using horseradish peroxidase. *J. Comp. Neurol.*, 182: 707–726, 1978.

64. Norden, J. J., Lin, C.-S., and Kaas, J. H. Subcortical projections of the dorsomedial visual area (DM) of visual association cortex in the owl monkey *Aotus trivirgatus. Exptl. Brain Res.*, 32: 321.334, 1978.

65. Ogren, M., and Hendrickson, A. Pathways between striate cortex and subcortical regions in *Macaca mulatta* and *Saimiri sciureus:* Evidence for a reciprocal pulvinar connection. *Exptl. Neurol.*, 53: 780–800, 1976.

66. Ogren, M. P., and Hendrickson, A. E. The distribution of pulvinar terminals in visual areas 17 and 18 of the monkey. *Brain Res.*, 137: 343–350, 1977.

67. Pandya, E. N., and Kuypers, H. G. Cortico-cortical connections in the rhesus monkey. *Brain Res.*, 13: 13–36, 1969.

68. Raczkowski, D., and Diamond, I. T. Connections of the striate cortex in *Galago senegalensis. Brain Res.*, 144: 383–388, 1978.

69. Rezak, M., and Benevento, L. A. A comparison of the organization of the projections of the dorsal lateral geniculate nucleus, the inferior pulvinar and adjacent lateral pulvinar to primary visual cortex (area 17) in the macaque monkey. *Brain Res.*, 167: 19–40, 1979.

70. Rocha-Miranda, C. E., Bender, D. B., Gross, C. G., and Mishkin, M. Visual activation of neurons in inferotemporal cortex depends on striate cortex and forebrain commisures. *J. Neurophysiol.*, 38: 475–491, 1975.

71. Rowe, M., and Stone, J. Naming of neurons. Classification and naming of cat retinal ganglion cells. *Brain, Behav. Evol.*, 14: 185–216, 1977.

72. Schiller, D. H., and Koerner, F. Discharge characteristics of single units in superior colliculus of the alert rhesus monkey. *J. Neurophysiol.*, 34: 920–936, 1971.

73. SCHILLER, H., AND MALPELI, J. G. Properties and tectal projections of monkey retinal ganglion cells. *J. Neurophysiol.*, 40: 428–445, 1977.
74. SCHILLER, P. H., AND MALPELI, J. G. Functional specificity of lateral geniculate nucleus laminae of the rhesus monkey. *J. Neurophysiol.*, 41: 788–797, 1978.
75. SCHILLER, P. H., AND STRYKER, M. Single-unit recording and stimulation in superior colliculus of the alert rhesus monkey. *J. Neurophysiol.*, 35: 915–924, 1971.
76. SELTZER, B., AND PANDYA, D. N. Afferent cortical connections and architectonics of the superior temporal sulcus and surrounding cortex in the rhesus monkey. *Brain Res.*, 149: 1–24, 1978.
77. SHERMAN, S., WILSON, J., KAAS, J., AND WEBB, S. X- and Y-cells in the dorsal lateral geniculate nucleus of the owl monkey *(Aotus trivirgatus)*. *Science*, 192: 475–477, 1976.
78. SPARKS, D. L. Functional properties of neurons in the monkey superior colliculus: coupling of neuronal activity and saccade onset. *Brain Res.*, 156: 1–16, 1978.
79. SPATZ, W. B. An efferent connection of the solitary cells of Meynert. A study with horseradish peroxidase in the marmoset, *Callithrix*. *Brain Res.*, 97: 450–455, 1975.
80. SPATZ, W. B. Topographically organized reciprocal connections between areas 17 and MT (visual area of superior temporal sulcus) in marmoset, *Callithrix jacchus*. *Exptl. Brain Res.*, 27: 91–108, 1977.
81. SPATZ, W. B., AND ERDMANN, G. Striate cortex projections to the lateral geniculate and other thalamic nuclei: a study using degeneration and autoradiographic tracing methods in the marmoset, *Callithrix*. *Brain Res.*, 82: 91–108, 1974.
82. SPATZ, W. B., AND TIGGES, J. Species difference between Old World and New World monkeys in the organization of the striate-prestriate association. *Brain Res.*, 43: 591–594, 1972a.
83. SPATZ, W. B., AND TIGGES, J. Experimental anatomical studies on the "Middle Temporal Visual Area (MT)" in primates. I. Efferent corticocortical connections in the marmoset, *Callithrix jacchus*. *J. Comp. Neurol.*, 146: 451–464, 1972b.
84. SPATZ, W. B., AND TIGGES, J. Studies on the visual area MT in primates. II. Projection fibers to subcortical structures. *Brain Res.*, 61: 371–378, 1973.
85. SPATZ, W. B.., TIGGES, J., AND TIGGES, M. Subcortical projections, cortical associations and some intrinsic interlaminar connections of the striate cortex in the squirrel monkey. *(Saimiri)*. *J. Comp. Neurol.*, 140: 155–174, 1970.
86. SYMONDS, L. L., AND KAAS, J. H. Connections of striate cortex in the prosimian, *Galago senegalensis*. *J. Comp. Neurol.*, 181: 477–512, 1978.
87. TIGGES, J., BOS, J., AND TIGGES, M. An autoradiographic investigation of the subcortical visual system in chimpanzees. *J. Comp. Neurol.*, 172: 367–380, 1977.
88. TIGGES, J., SPATZ, W. B., AND TIGGES, M. Reciprocal point-to-point

connections between parastriate and striate cortex in the squirrel monkey. *(Saimiri). J. Comp. Neurol.*, 148: 481–490, 1973a.

89. TIGGES, J., SPATZ, W. B., AND TIGGES, M. Efferent cortico-cortical fiber connections of area 18 in the squirrel monkey *(Samiri). J. Comp. Neurol.*, 158: 219–236, 1974.

90. TIGGES, J., TIGGES, M., AND KALAHA, C. S. Efferent connections of area 17 in *Galago. Amer. J. Physical Anthro.*, 38: 393–398, 1973b.

91. TIGGES, J., TIGGES, M., AND PERACHIO, A. Complementary laminar terminations of afferents to area 17 originating in area 18 and in the lateral geniculate nucleus in squirrel monkey. *J. Comp. Neurol.*, 176: 371–396, 1976.

92. TROJANOWSKI, J. Q., AND JACOBSON, S. Areal and laminar distribution of some pulvinar cortical efferents in rhesus monkey. *J. Comp. Neurol.*, 169: 371–396, 1976.

93. TROJANOWSKI, J. Q., AND JACOBSON, S. The morphology and laminar distribution of cortico-pulvinar neurons in the rhesus monkey. *Exptl. Brain Res.*, 28: 51–62, 1977.

94. UNGERLEIDER, L. G., AND MISHKIN, M. The striate projection zone in the superior temporal sulcus of *Macaca mulatta:* location and topographic organization. *J. Comp. Neurol.*, 188: 347–366, 1979.

95. VAN ESSEN, D. C., MAUNSELL, J. H. R., AND BIXBY, J. L. The organization of extrastriate visual areas in the macaque monkey. This volume, chapter 14.

96. VAN ESSEN, D. C., AND ZEKI, S. M. The topographic organization of prestriate cortex. *J. Physiol., London,* 277: 193–226, 1978.

97. WAGOR, E., LIN, C.-S., AND KAAS, J. H. Some cortical projections of the dorsomedial visual area (DM) of association cortex in the owl monkey, *Aotus trivirgatus. J. Comp. Neurol.*, 163: 227–250, 1975.

98. WALL, J. T., SYMONDS, L. L., AND KAAS, J. H. Projections of the middle temporal visual area (MT) in the prosimian *Galago senegalensis. Arvo Abst.*, 1, 1980.

99. WELLER, R. E., AND KAAS, J. H. Connections of striate cortex with the posterior bank of the superior temporal sulcus in macaque monkeys. *Neurosci. Abst.*, 4: 650, 1978.

100. WELLER, R. E., AND KAAS, J. H. Connections of the dorsolateral visual area (DL) of extrastriate visual cortex of the owl monkey *(Aotus trivirgatus) Neurosci. Abst.*, 6: 579, 1980.

101. WELLER, R. E., KAAS, J. H., AND WETZEL, A. B. Evidence for the loss of X-cells of the retina after long-term ablation of visual cortex in monkeys. *Brain Res.*, 160: 134–138, 1979.

102. WELLER, R. E., GRAHAM, J., AND KAAS, J. H. Cortical connections of striate cortex in macaque monkeys. *ARVO Abst.*, 157, 1979.

103. WHITTERIDGE, D. Area 18 and the vertical meridian of the visual field. In: *Functions of the Corpus Callosum,* edited by E. G. ETTLINGER, London: Ciba Foundation, 1965, pp. 115–120.

104. WILSON, M., AND TOYNE, N. Retino-tectal and cortico-tectal projections in *Macaca mulatta. Brain Res.*, 24: 395–406, 1970.

105. WONG-RILEY, M. T. T. Demonstration of the geniculo-cortical and callosal projection neurons in the squirrel monkey by means of retrograde axonal transport of horseradish peroxidase. *Brain Res.*, 79: 267–272, 1974.

106. WONG-RILEY, M. Autoradiographic studies of subcortical and cortical projections from striate and extrastriate cortices of squirrel and macaque monkeys. *Anat. Rec.*, 184: 566, 1976.

107. WONG-RILEY, M. T. T. Connections between the pulvinar nucleus and the prestriate cortex in the squirrel monkey as revealed by peroxidase histochemistry and autoradiography. *Brain Res.*, 134: 249–267, 1977.

108. WONG-RILEY, M. Reciprocal connections between striate and prestriate cortex in squirrel monkey as demonstrated by combined peroxidase histochemistry and autoradiography. *Brain Res.*, 147: 159–164, 1978.

109. WONG-RILEY, M. Columnar cortico-cortical interconnections within the visual system of the squirrel and macaque monkeys. *Brain Res.*, 162: 201–217, 1979.

110. WURTZ, R. H., AND GOLDBERG, M. E. Activity of superior colliculus in behaving monkey. IV. Effects of lesions on eye movements. *J. Neurophysiol.*, 35: 587–596, 1972.

111. ZEKI, S. M. Representation of central visual fields in prestriate cortex of monkey. *Brain Res.*, 14: 271–291, 1969.

112. ZEKI, S. M. Interhemispheric connections of prestriate cortex in monkey. *Brain Res.*, 19: 63–75, 1970.

113. ZEKI, S. M. Convergent input from the striate cortex (area 17) to the cortex of the superior temporal sulcus in the rhesus monkey. *Brain Res.*, 28: 338–340, 1971a.

114. ZEKI, S. M. Cortical projections from two prestriate areas in the monkey. *Brain Res.*, 34: 19–35, 1971b.

115. ZEKI, S. M. The projections to the superior temporal sulcus from areas 17 and 18 in the rhesus monkey. *Proc. Roy. Soc., London, B,* 193: 199–207, 1976.

116. ZEKI, S. M. The cortical projections of foveal striate cortex in the rhesus monkey. *J. Physiol., London,* 227–244, 1978a.

117. ZEKI, S. M. The third visual complex of rhesus monkey prestriate cortex. *J. Physiol., London,* 277: 245–272, 1978b.

118. ZEKI, S. M. Functional specialization in the visual cortex of the rhesus monkey. *Nature,* 274: 423–428, 1978c.

119. ZEKI, S. M. AND SANDEMAN, D. R. Combined anatomical and electrophysiological studies on the boundary between the second and third visual areas of rhesus monkey cortex. *Proc. Roy. Soc., London, B,* 194: 555–562, 1976.

Chapter 6

Organization of Extrastriate Visual Areas in the Macaque Monkey

D. C. Van Essen, J. H. R. Maunsell and J. L. Bixby

Division of Biology, California Institute of Technology, Pasadena, California

1. Introduction

In primates, the regions of cerebral cortex that are specifically visual in function occupy the entire occipital lobe plus substantial portions of the temporal and parietal lobes. The division of the occipital lobe into three areas by Brodmann (4) and others on the basis of cytoarchitectonic criteria has been widely accepted, but recent studies have demonstrated that the extrastriate occipital cortex (i.e., Brodmann's areas 18 and 19) actually contains numerous distinct visual areas. In an Old World monkey, the macaque, five such areas have been identified to date, and in a New World monkey, the owl monkey, seven extrastriate areas have been found in the occipital and temporal lobes (1, 14, 21).

We, along with many others, have chosen the macaque as an experimental animal because of several significant advatages it offers for the study of visual function. These include excellent visual acuity, stereopsis and color vision; moreover, the macaque ranks high among laboratory animals in terms of intelligence, amenability to behavioral studies and phylogenetic proximity to humans. It has a major drawback, however, which is the extensive folding of the cortical surface. Much of the cortex is buried within sulci, and in the occipital lobe these convolutions are particularly deep and complex. The resultant variations in the shapes of histological sections taken at different levels through the cortex have posed formidable problems for analyzing and interpreting experiments on visual cortical organization.

One way to circumvent the problems imposed by cortical convolutions is to devise a method of "unfolding" the cortex and representing its surface on a two-dimensional sheet. This has often been done in a purely schematic fashion, but it is obviously preferable to have a more accurate mapping procedure that could be applied to any particular hemisphere and which would minimally distort the representation of cortical surface area. Such a method was recently developed and applied to the analysis of interhemispheric connections in part of the macaque occipital lobe (17). We have now extended this approach so that the entire cerebral neocortex can be mapped onto a single two-dimensional surface (15).

2. Two-Dimensional Cortical Maps

Figure 6.1 shows a cortical map that was made from the series of horizontal sections illustrated in Brodmann's manuscript on cytoarchitectonic areas in Old World monkeys (4). Each of the numbered contour lines in the map represents the outline of a particular section, whose dorsoventral level is indicated on the drawings of the hemisphere. The stippled regions on the map indicate cortex that is buried within sulci and the dashed lines mark the location of the fundus of each sulcus. Along the perimeter of the map, the dotted lines indicate the border of neocortex with various allocortical structures. The remainder of the perimeter (solid lines) reflects discontinuities in the representation that were made to reduce spatial distortions—just as cartographers make cuts in the representation of the earth's surface in order to minimize distortions. For the macaque brain, it is possible to make a reasonably undistorted map

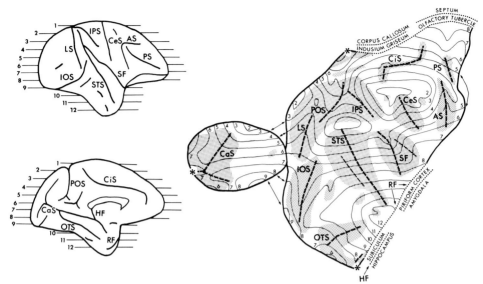

FIG. 6.1. A two-dimensional map of cerebral neocortex in an Old World monkey. Each contour in the map represents the outline of layer IV in a particular section (altered in shape but not in length). Stippled regions: cortex within sulci; dashed lines: fundi of sulci. Abbreviations; AS, arcuate sulcus; CaS, calcarine sulcus; CeS, central sulcus; CiS, cingulate sulcus; HF, hippocampal fissure; IOS, inferior occipital sulcus; IPS, intraparietal sulcus; LS, lunate sulcus; RF, rhinal fissure; SF, Sylvian fissure; STS, superior temporal sulcus. Reproduced with permission from the *Journal of Comparative Neurology.*

with a long cut along the border of striate cortex and a shorter cut in the frontal lobe, between the arcuate and principal sulci.

The principle involved in generating cortical maps of the type shown in Fig. 6.1 is to arrange that the contours representing adjacent sections be evenly spaced and appropriately aligned with respect to one another. This can be done by changing the shape, but not the length, of each contour as it is transferred from the section (where it represents the outline of layer IV) to the cortical map. By trial and error the contours are adjusted to minimize distortions, i.e., to insure that neighboring points on the map everywhere correspond to neighboring sites within the cortex. Comparisons with a computer-generated, three-dimensional reconstruction of the occipital lobe indicate that surface areas of most regions are represented with reasonable accuracy (i.e., ± 20%) on cortical maps (15). Planimetric measurements on these maps can thus yield useful information: for instance, in the macaque, the striate cortex occupies about one-sixth of the total neocortex; about half of the cortex is buried within sulci.

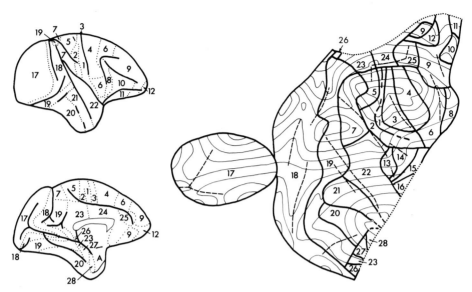

Fig. 6.2. Cytoarchitectonic areas of Brodmann shown on the cortical map. In some regions, precise borders between areas were not described (e.g., areas 1 and 2); these uncertainties are indicated by heavy
dashed lines. Fine dashed lines demarcate sulcal fundi, as in Fig. 6.1. Note
that in Brodmann's scheme area 18 surrounds area 17 completely,
whereas there is a small gap in area 19 within the calcarine sulcus. Reproduced with permission from the *Journal of Comparative Neurology.*

The borders of Brodmann's cytoarchitectonic areas are plotted
on the cortical map shown in Fig. 6.2. Area 18 is by far the largest
subdivision described by Brodmann. Thus, it perhaps should not
be surprising that neither area 18 nor, for that matter, area 19
turns out to be a single functional entity.

3. Visual Areas of the Occipital Lobe

The evidence for multiple visual areas within areas 18 and 19
comes mainly from experiments that directly or indirectly reveal the
topographic organization of the cortex, i.e., the way in which the
visual hemifield is represented on the cortical surface. This information has been obtained by (*a*) physiological mapping of receptive
fields, (*b*) anatomical tracing of pathways from areas whose visual
topography is already known, and (*c*) mapping of interhemispheric
connections as an indication of where the vertical meridian is represented. These approaches have together demonstrated the existence of at least five topographically organized visual areas, V2(V II),

V3(V III), V3A(V IIIA), V4(V IV) and MT (7, 8, 16–19). It has proven difficult, however, to determine the full extent and precise boundaries of any of these areas. This is partly because of the aforementioned problems of dealing with a convoluted hemisphere, but in addition there are difficulties associated with intrinsic complexities of visual topography in individual extrastriate areas (see below).

3.1. The Location of Area MT

One useful approach to this problem has been to examine cortical architecture in experiments for which independent evidence on the location of visual areas is available. We have found that the architecture of extrastriate cortex in the occipital and temporal lobes shows numerous regional variations in sections stained for myelin, even though adjacent Nissl-stained sections appear relatively uniform in cytoarchitecture. One of the clearest myeloarchitectonic subdivisions is situated in the posterior bank of the superior temporal sulcus (STS), in a region known to receive a direct input from striate cortex (6, 18). Similarities in location, connections and function indicate that the striate-receptive area in the macaque STS is almost certainly homologous with the middle temporal area (MT) of New World primates (2), and we therefore refer to it by that name (see ref. 14, 16).

The myeloarchitecture of the posterior bank of the STS is illustrated in Fig. 6.3, which shows a photomicrograph of a section cut in the parasagittal plane. The dorsal part of the sulcus is characteristically rather lightly myelinated, and it is usually possible to recognize distinct bands of myelinated fibers in layers IV and Vb, the outer and inner bands of Baillarger, respectively. Immediately ventral to this region is a heavily myelinated zone in which the two bands of Baillarger are not recognizable. Further ventrally, the myelination is intermediate in density relative to the other two zones.

The location and overall extent of the heavily myelinated zone on the posterior bank of the STS is indicated in the cortical map of Fig. 6.4. It is evident that the heavily myelinated area (solid outline) occupies only a small portion of the STS. The relationship of this zone to visual area MT was examined in the same animal by making multiple lesions on the exposed surface of striate cortex (Fig. 6.4, left). The resultant zone of degeneration was restricted to the lower part of the heavily myelinated zone (Fig. 6.4, right). In a separate experiment [^{3}H]-proline was injected into striate cortex within the calcarine sulcus, where peripheral visual fields are represented; the transported label in the STS was restricted to the upper part of the

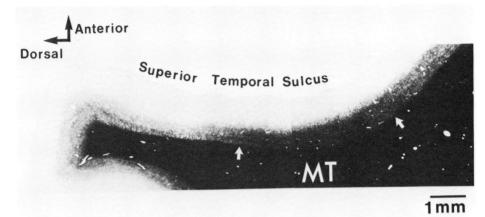

FIG. 6.3. Myeloarchitecture of the posterior bank of the STS. The section was cut in the parasagittal plane several mm lateral to the fundus of the sulcus. Borders of the heavily myelinated zone are indicated by arrows.

heavily myelinated zone, i.e., the part in which degeneration was absent in Fig. 6.4. We conclude, therefore, that the heavily myelinated zone is precisely coextensive with MT, as defined by striate cortical projections. These results also confirm that at least a crude topographic organization exists within MT, with central fields represented ventrally and peripheral fields dorsally in MT (7, 12, 13, 16; see below).

The size, shape and location of MT is remarkably consistent in the 12 heispheres for which we have complete maps. MT is approximately elliptical in shape, with a long axis of 10 mm on average and a short axis of 3.5 mm. Its orientation within the hemisphere is illustrated in Fig. 6.5, which shows a face-on view of the posterior bank of the STS, after dissecting away the anterior part of the brain. The long axis of MT runs obliquely from dorso-medio-posterior to ventro-latero-anterior. It almost always terminates within a small secondary fold of the STS, thus providing a convenient geographical basis for estimating the location of MT. The total surface area of MT determined from planimetric measurements on cortical maps is 30 mm² (range, 25–35 mm²); this is less than 3% of the area of striate cortex.

3.2. Functional and Topographic Organization of MT

An important functional characteristic of MT is the high incidence of cells displaying a marked selectivity for visual stimuli moving in particular directions (20, 21). The availability of myeloarchitectonic criteria for identifying the borders of MT made it feasible to

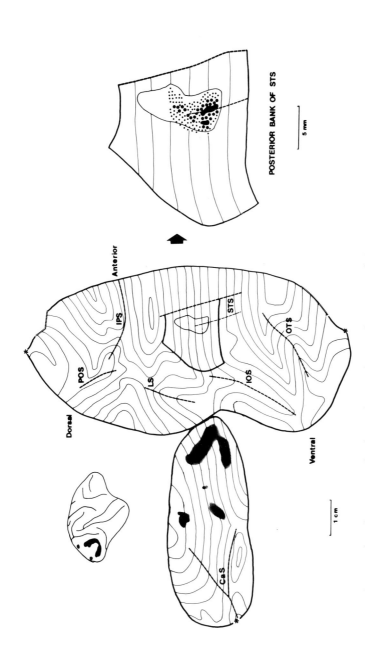

FIG. 6.4. Relationship between the heavily myelinated region of the STS and visual area MT as indicated by striate cortical projections. Lesions of striate cortex are indicated by darkened areas (left); degeneration within MT is shown in the expanded view of the posterior bank of the STS (right). Small and large dots indicate sparse and moderate levels of degeneration, respectively. Note that the degeneration is contained almost entirely within the heavily myelinated zone (solid line). Reproduced with permission from the *Journal of Comparative Neurology.*

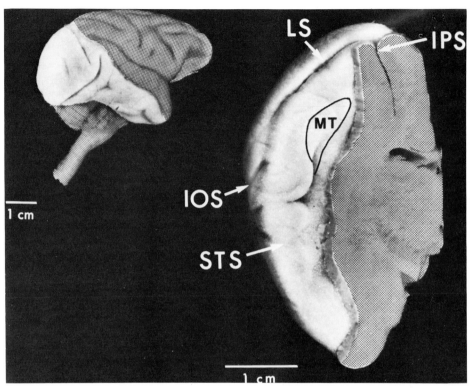

Fɪɢ.   6.5.   The location of MT within the STS. Shaded portions of the brain shown in the upper left were removed by dissection in order to expose the posterior bank of the STS for a face-on view (right). Shaded region on right indicates cut white matter and subcortical regions. Solid line indicates the presumed location of MT, based on its consistent location in all hemispheres that have been examined histologically. Reproduced with permission from the *Journal of Comparative Neurology.*

determine the extent to which directionally selective cells are restricted specifically to MT. Fig. 6.6 shows the results of a physiological experiment in which four parallel microelectrode penetrations were made into the STS. Three of the penetrations entered MT after passing through cortex adjoining its dorsolateral border. The responses to moving visual stimuli evoked at each recording site were characterized as strongly direction selective (filled circles), moderately direction selective (crosses), or direction nonselective (open triangles). Almost all of the responses recorded within MT showed strong or moderate direction selectivity, and the sharpness of the functional transition along the dorsolateral border matched the precision with which we could determine this border anatomically.

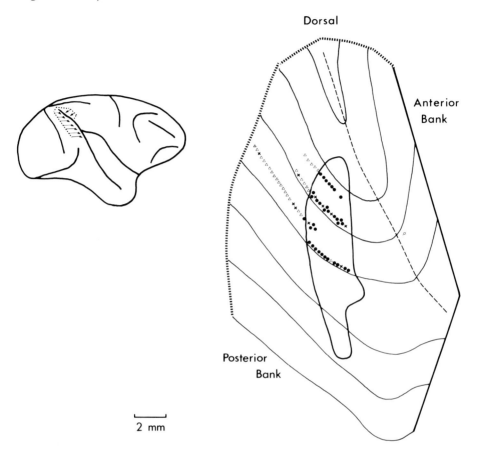

Dorsal

Anterior
Bank

Posterior
Bank

2 mm

SUPERIOR TEMPORAL SULCUS

FIG. 6.6. Functional properties of cells in the posterior bank of the STS. Receptive fields of single cells or multicell clusters were examined at 0.25 mm intervals in penetrations with low-resistance tungsten microelectrodes. Responses were classified as direction nonselective (open triangles), moderately direction selective (crosses) or strongly directional selective (filled circles). Myeloarchitectonic border of MT is indicated by solid line. Reproduced with permission from the *Journal of Comparative Neurology.*

Markedly different results were found in recordings traversing the medial border of MT. Most of the responses in the region medial to MT showed direction selectivity as striking as that within MT; the principal difference we noted was the very large size of receptive fields in the cortex medial to MT. Thus, there is evidently more than one visual area in the STS concerned with the analysis of moving stimuli. Whether this additional area is part of the polysensory zone described on the anterior banks of the STS (5) has not been determined.

Given the apparent homology of MT in Old World and New World primates, one might expect that MT in the macaque would contain a simple point-to-point, or "first-order" transformation of the visual hemifield, as has been described for MT in the owl monkey (2) and bushbaby (3). Suggestions that this is, indeed, the case have been made on the basis of the organization of inputs from striate cortex (7, 12, 13). On the other hand, Zeki (21) and Montero (10) have suggested that MT has a complex topographic organization, and our own experiments support this view. One line of evidence comes from anatomical tracing of the projections of striate cortex, using combined lesions and [^{3}H]-proline injections. In general, the pattern of degeneration and transported label in MT is not simply a reduced version of the pattern of lesions and injection sites, as would be expected from a compressed, but otherwise undistorted, mapping of striate cortex onto MT. In one instance, a single injection of [^{3}H]-proline led to two discrete patches of transported label within MT. A second line of evidence comes from the receptive field locations determined in physiological recordings. In a first-order transformation, the inferior and superior vertical meridians should be represented along opposite borders of MT; although this is indeed the case in some of our penetrations through MT, several clear exceptions have been observed. A third, less direct line of evidence comes from the pattern of interhemispheric connections, which in other visual areas reflects the representation of the vertical meridian of the visual field (17). In the bushbaby and owl monkey, the callosal inputs to MT are densest along its perimeter (11), as would be anticipated from its known topographic organization (2, 3). In the macaque, however, we have found that the callosal inputs to MT are patchy, irregular and not specifically concentrated along its perimeter. Thus, it seems likely that the internal organization of MT is not identical in all species and that its topographic organization in the macaque is complex and not fully understood. It will be interesting to determine whether there are any basic similarities to the complex, but highly precise, somatotopic patterns recently found in primate somatosensory cortex (9).

3.3. Interhemispheric Connections

A final topic to consider briefly is the organization of other extrastriate areas besides MT. One useful approach to understanding the overall layout of visual areas is to map the distribution of interhemispheric connections, which, as mentioned above, provide an indication of where the vertical meridian is represented in each area. In a previous study (17), a complex but consistent pat-

tern of callosal inputs to the dorsal half of the occipital lobe was described. In extending this map to the remainder of the occipital lobe
plus portions of the temporal and parietal lobes, we have discovered
a striking asymmetry in the organization of dorsal and ventral
halves of the hemisphere (Fig. 6.7). The most obvious aspect of this
asymmetry is the simplicity of the pattern ventrally relative to that
dorsally. Anterior to the narrow strip of callosal inputs along the

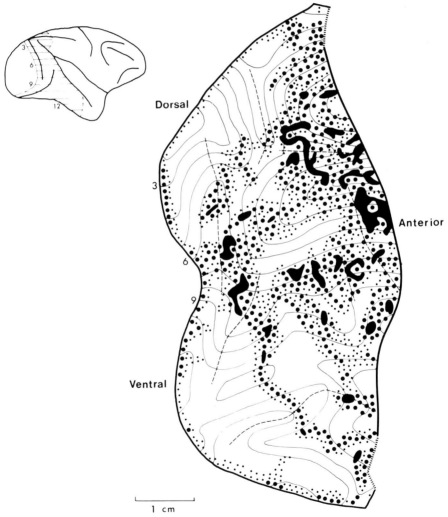

FIG. 6.7. Interhemispheric connections in extrastriate visual cortex
of the occipital, temporal and parietal lobes. The cortical map represents
the portion of the hemisphere indicated by horizontal lines in the drawing
(upper left). Degeneration following transection of the corpus callosum
was classified as dense (solid areas), moderate (large dots), or sparse
(small dots).

boundary of striate cortex, there is a sharply defined band running the full width of the hemisphere ventrally, from the lateral margin of the inferior occipital sulcus to the medial margin of the occipito-temporal sulcus. We have anatomical evidence (unpublished experiments) that this band, where the vertical meridian presumably is represented, lies well anterior to the border of V2(V II), where the horizontal meridian is represented. Whether the intervening region of cortex contains only a single visual representation is currently being examined. In this regard, it is interesting to note that a similar pattern of ventral interhemispheric connections has been found in extrastriate cortex of the owl monkey (11), where there is evidence for only a single visual area adjoining ventral V2(V II) (see chapter 7 by Allman).

4. Conclusion

The major development in our understanding of extrastriate visual cortex over the past decade has been the realization that in higher mammals there are numerous areas that vary in size, internal organization and pattern of connections. Elucidation of the specific functions of individual areas has lagged somewhat, largely because a reliable definition of areal boundaries is a prerequisite to detailed functional studies. Nonetheless, there are clear indications of major differences in the types of visual processing in different areas (cf. 14, 21). As the basic layout of extrastriate visual areas becomes progressively better understood, it is reasonable to anticipate accelerated progress in clarifying their roles in visual perception.

Acknowledgments

We thank P. Knudsen for excellent assistance with histology and in preparation of figures. This research was supported by NIH grant R01 EY02091 from the National Eye Institute and by a grant from the Sloan Foundation.

References

1. ALLMAN, J. M. Evolution of the visual system in the early primates. *Prog. Psychobiol. Physiol. Psychol.*, 7: 1–53, 1977.
2. ALLMAN, J. M., AND KAAS, J. H. A representation of the visual field in the caudal third of the middle temporal gyrus of the owl monkey (*Aotus trivirgatus*). *Brain Res.*, 31: 85–105, 1971.

3. ALLMAN, J. M., KAAS, J. H., AND LANE, R. H. The middle temporal visual area (MT) in the bushbaby, *Galago senegalensis. Brain Res.,* 57: 197–202, 1973.

4. BRODMANN, K. Beitrage zur histologischen Localisation der Gross-hirnrinde. Dritte Mitteilung. Die Rindenfelder der nierderen Affen. *J. Psychol. Neurol., Leipzig,* 4: 177–226, 1905.

5. BRUCE, C. J., DESIMONE, R., AND GROSS, C. G. Large visual recep-tive fields in a polysensory area in the superior temporal sulcus of the macaque. *Soc. Neurosci. Abstr.,* 3: 554, 1977.

6. CRAGG, B. G. The topography of the afferent projections in circum-striate visual cortex of the monkey studied by the Nauta method. *Vi-sion Res.,* 9: 733–747, 1969.

7. GATTASS, R., AND GROSS, C. G. A visuotopically organized area in the posterior superior temporal sulcus of the macaque. *Assoc. Res. Vision Opthalmol.,* 18: 184 (Abstr.), 1979.

8. MAUNSELL, J. H. R., BIXBY, J. L., AND VAN ESSEN, D. C. The middle temporal area (MT) in the macaque: architecture, functional proper-ties and topographic organization. *Soc. Neurosci Abstr.,* 5: 796, 1979.

9. MERZENICH, M. M., KAAS, J. H. SUR, M., AND LIN, C.-S. Double rep-resentation of the body surface within cytoarchitectonic areas 3b and 1 in "SI" in the owl monkey (*Aotus trivirgatus*). *J. Comp. Neurol.,* 181: 47–74, 1978.

10. MONTERO, V. M. Patterns of connections from the striate cortex to cortical visual areas in superior temporal sulcus of macaque and mid-dle temporal gyrus of owl monkey. *J. Comp. Neurol.* 189: 45–55, 1980.

11. NEWSOME, W. T., AND ALLMAN, J. M. The interhemispheric connec-tions of visual cortex in the owl monkey, *Aotus trivirgatus,* and the bushbaby, *Galago senegalensis. J. Comp. Neurol.,* 194: 209–234, 1980.

12. UNGERLEIDER, L. G., AND MISHKIN, M. The visual area in the supe-rior temporal sulcus of *Macaca mulatta;* location and topographic or-ganization. *Anat. Rec.,* 190: 568, 1978.

13. WELLER, R. E., AND KAAS, J. H. Connections of striate cortex with the posterior bank of the superior temporal sulcus in macaque mon-keys. *Soc. Neurosci. Abstr.,* 4: 650, 1978.

14. VAN ESSEN, D. C. Visual areas of the mammalian cerebral cortex. *Ann. Rev. Neurosci.,* 2: 227–263, 1979.

15. VAN ESSEN, D. C., AND MAUNSELL, J. H. R. Two-dimensional maps of the cerebral cortex. *J. Comp. Neurol.* 191: 255–281, 1980.

16. VAN ESSEN, D. C., MAUNSELL, J. H. R., AND BIXBY, J. L. The middle temporal visual area in the macaque: myeloarchitecture, connec-tions, functional properties and topographic organization, *J. Comp. Neurol.,* 199: 293–326.

17. VAN ESSEN, D. C., AND ZEKI, S. M. The topographic organization of rhesus monkey prestriate cortex. *J. Physiol., London,* 277: 193–226, 1978.

18. ZEKI, S. M. Representation of central visual fields in prestriate cortex of monkey. *Brain Res.*, 14: 271–291, 1969.
19. ZEKI, S. M. Cortical projections from two prestriate areas in the monkey. *Brain Res.*, 34: 19–35, 1971.
20. ZEKI, S. M. Functional organization of a visual area in the posterior bank of the superior temporal sulcus of the rhesus monkey. *J. Physiol., London*, 549–573, 1974.
21. ZEKI, S. M. Functional specialization in the visual cortex of the rhesus monkey. *Nature*, 274: 423–428, 1978.

Chapter 7

Visual Topography and Function

Cortical Visual Areas in the Owl Monkey

John M. Allman, James F. Baker, William T. Newsome and Steven E. Petersen

Division of Biology, California Institute of Technology, Pasadena, CA 91125

1. Topographic Organization

The functional division of labor among the large number of visual areas in the cerebral cortex in primates constitutes one of the great scientific riddles in biology. The topographic organization of many of these areas was first mapped in the owl monkey (2–7, 39). The cortical visual areas of the owl monkey are illustrated in an unfolded schema in Fig. 7.1 and as they are located in the brain in Fig.

171

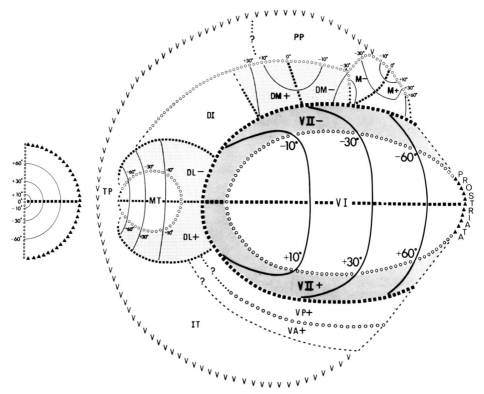

Fig. 7.1. A schematic unfolding of the visual cortex of the left hemisphere of the owl monkey. The visual cortex corresponds to approximately the posterior third of the entire neocortex. The unfolded visual cortex is approximately a hemispherical surface, which is viewed from above in this diagram. The perimeter chart on the left shows the contralateral (right) half of the visual field. The symbols in this chart are superimposed on the surface of the visual cortex. Pluses indicate upper quadrant representations; minuses, lower quadrants; dashed lines, borders of areas that correspond to the representation of the relatively peripheral parts of the visual field, but not to the extreme periphery. The row of Vs indicates the approximate anterior border of visually responsive cortex. The dotted lines broken by question marks indicate uncertain borders. DI, Dorsointermediate Visual Area; DL, Dorsolateral Crescent Visual Area; DM, Dorsomedial Visual Area; IT, inferotemporal Cortex; M, Medial Visual Area; MT, Middle Temporal Visual Area, PP, Posterior Parietal Cortex; VA, Ventral Anterior Visual Area; VP, Ventral Posterior Visual Area.

7.2. Areas V I and V II share a common border, along which the vertical meridian or midline of the visual field is represented in each area. In each hemisphere V I contains a topological, first-order transformation, of the contralateral half of the visual field, in which the more central portions are greatly expanded (3). In V II, the repre-

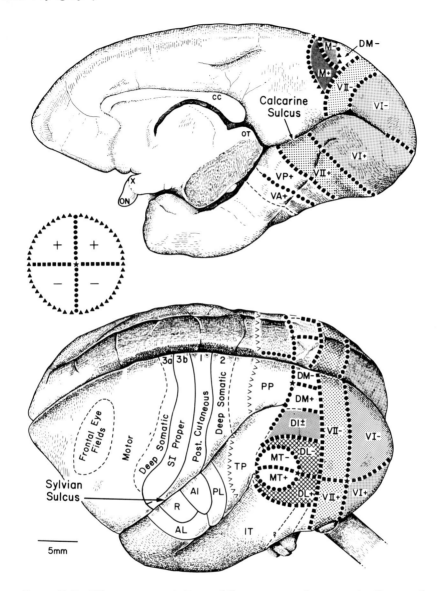

FIG. 7.2. The representation of the sensory domains in the cerebral cortex of the owl monkey. Above is a ventromedial view of the right hemisphere; below is a dorsolateral view. First Auditory Area; AL, Anterolateral Auditory Area; CC, Corpus Callosum; ON, Optic Nerve, OT, Optic Tectum; PL, Posterolateral Auditory Areas; R, Rostral Auditory Area. The cortical visual areas were mapped by Allman and Kaas (2–7) and Newsome and Allman (39); the somatosensory areas were mapped by Merzenich et al. (36); the auditory areas were mapped by Imig et al. (27). Other conventions and abbreviations are as in Fig. 7.1. Reproduced from reference 10 by permission of the American Physiological Society

sentation of the horizontal meridian is split except near the center of gaze (4). Since adjacent parts of the contralateral hemifield are not necessarily mapped onto adjacent parts of V II, the representation of the hemifield is not topological and has been termed a second order transformation (4). The Middle Temporal Area (MT) and the Dorsolateral Area (DL) form a miniature mirror-image of V I and V II (5). DL, together with the Medial (M), the Dorsointermediate (DI), the Dorsomedial (DM) and the Ventroposterior (VP) Areas, constitute a third tier of cortical visual areas that adjoin the anterior border of V II (5, 6, 7, 39) (V I is the first tier; V II the second). Two visuotopically organized fourth tier areas are known: MT and the recently discovered Ventral Anterior Area (VA) (2, 39). Each of these shares a border corresponding to a vertical meridian representation with a third tier area. The possibility exists that additional fourth tier areas will be discovered in the visually responsive posterior parietal (PP) or temporal parietal (TP) cortex that will be the mates to adjacent third tier areas. Finally, there is a large region of visually responsive cortex in the inferior temporal gyrus (IT) anterior to VA and DL. Only the posterior third of this region has been explored in the owl monkey and this zone contains neurons with receptive fields predominantly in the central visual field and with no apparent visuotopic organization (1). These results are similar to those obtained by Desimone and Gross (15) over a much broader region of the inferior temporal gyrus in macaque monkeys.

Several principles emerge from the topographic organization of the cortical visual areas.

1. A surprisingly large number of visuotopically organized areas exists—at least nine.

2. Since each of these exists as an anatomical entity, it is probable that each is a functional entity as well.

3. The areas group themselves into a number of larger (and not mutually exclusive) sets. These include the dyads that have adjoining vertical meridian representations: V I–V II, MT–DL and VP–VA. Other groupings include: the first-order transformations (VI and MT); the second order transformations (V II, DL, DM, M); the third tier (M, DM, DI, DL and VP); and the fourth tier (MT and VA). The areas in these sets may have important functional or developmental attributes in common.

4. The areas are juxtaposed so that their common borders correspond nearly always to portions of either the horizontal or the vertical meridian and the adjacent points on opposite sides of each border nearly always have similarly positioned receptive fields. Thus, there is very little disruption of visuotopic order at the interfaces between areas.

5. The relative representation of different portions of the visual field varies greatly among areas. Area V I has been widely assumed to contain a representation proportional to retinal ganglion cell density; however, recent analysis has shown that the representation of the central visual field in V I is very much greater relative to the periphery than could be accounted for on the basis of a proportional relationship (38). Each area has its own unique way of mapping the visual field and this mapping is likely to reflect the functions performed by that area (see Fig. 7.3). The representation of the central 10° of the visual field occupies 31% of the surface area of V I as compared to 75% of DL, which contains proportionally the largest expansion of the central visual field of all the areas (5). Only 4% is devoted to the central 10° of M, where the more peripheral parts of the visual field are relatively much better represented than in any other area (7).

6. Finally, there exists an important principle of topographic organization that cannot be deduced from these figures. Cortical visual areas differ in their visuotopic orderliness. Although we have not developed a direct quantitative measure of topographic orderliness, our experiences suggest that V I is the most orderly, followed

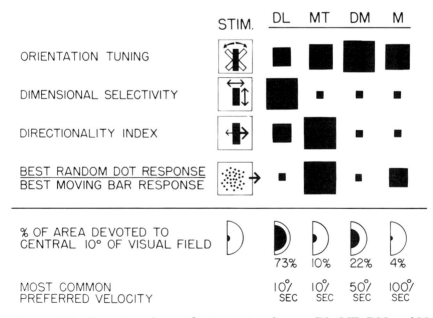

FIG. 7.3. Functional specificity in visual areas DL, MT, DM and M in the owl monkey. The strength of the functional attribute is indicated by the size of the black squares.

by V II, then by VP and VA, then by MT, DM and M. DL appears to be the least orderly of the mapped areas. One measure of the topographic order in these areas is the distribution of callosal terminals following section of the corpus callosum, which contains fibers that unify the two halves of the visual field along the vertical meridian (midline). The callosal terminals are largely restricted to a narrow zone on each side of the common border between V I and V II, along which the vertical meridian (midline) is represented (39). Along the border between VP and VA, callosal terminals are confined to a band that is nearly as discrete as that found at the V I–V II border and also corresponds to the representation of the vertical meridian (39). Callosal terminals form more diffuse zones of degeneration near the vertical meridian representations in MT, DM and M (39). Callosal terminals are distributed throughout most of DL, which contains large receptive fields that often overlap the vertical meridian (39, 42). Posterior Parietal (PP) and Temporal Parietal (TP) cortex may contain visutotopically organized areas, although it is difficult to discern a pattern, since receptive fields tend to be large and callosal connections diffuse in these regions. Finally, receptive fields in inferotemporal cortex (IT) tend to be located in the central visual field but do not appear to have a visuotopic organization (1). What organizing principle does prevail in this large expanse of visual cortex in the inferior temporal gyrus remains a tantalizing question. The topographic organization of IT may be related to its projections to perirhinal cortex and the amygdala and the related functions of visual memory and visual perception of emotionally significant stimuli (52).

2. Functional Correlates

Our guiding hypothesis has been that each cortical representation of the visual field performs its own set of neural functions, some of which are distinctive to that particular area. Our approach to determining the functions of cortical visual areas has been to analyze quantitatively the response properties of single neurons in each area in order to discover how information is processed in that area.

   The procedures employed for recording from single neurons in the cortical visual areas of owl monkeys are described in detail elsewhere (10, 42). Briefly, in order to make more efficient use of our owl monkeys, we have developed a chronic technique for recording repeatedly from the visual cortex in monkeys that have been tranquilized with light doses of triflupromazine and ketamine and their locally anesthetized eyes held stable by rings machined to fit the contour of the eyes. The visuotopic organization of the region of exposed cortex beneath the chronically implanted recording chamber was mapped in order to identify the location and boundaries of visual areas, so that recorded single neurons could be assigned to a particular area. Small electrolytic lesions were made at the end of selected microelectrode penetrations, identified in histogical sections and compared with the known cyto- and myeloarchitecture of the cortical visual area (2–6). Visual stimuli were light and dark bars, spots and textures focused on a tangent screen by a rear projection optic stimulator. After receptive field mapping and initial qualitative examination, the response properties were studied quantitatively using the optic stimulator in conjunction with a Nova 2 computer. For each stimulus parameter (orientation, direction of movement, etc.), the stimuli were presented in pseudorandom order so as to average out any periodic waxing and waning in a cell's responsiveness and to avoid possible habituating effects that may occur when the same stimulus is presented repeatedly (24). The computer recorded the neuron's spontaneous firing during an interval preceding the stimulus presentation and calculated its response relative to the level of spontaneous activity. From these data, the computer calculated a series of indices that enabled us to compare objectively the response profiles of the populations of neurons obtained from MT, DL, DM and M for orientation tuning and selectivity for direction of movement and stimulus dimensions. The following is a summary of these results (10, 42).

1. Neurons in DL, MT, DM and M are orientation selective. As in V I (36), neurons in these four extrastriate areas tend to be orientation selective, when tested with stationary flashed or moving bars; however, there are statistically significant differences among the areas in sharpness of tuning for orientation with DM neurons being the most selective, MT and M neurons intermediate and DL neurons least tuned (10).

2. In DL, about 70% of the neurons are selective for the spatial dimensions of visual stimuli within excitatory receptive fields that are generally much larger than the preferred stimulus dimensions. The dimensional selectivity of DL cells is independent of the sign of contrast in the receptive field (equal to light-on-dark and dark-on-light stimuli), the amount of contrast (similar response over a 1.5 log unit change in intensity) and the position of the stimulus within the receptive field. DL neurons have a wide range of preferred sizes, from 1 to 30° in length and from 0.25 to 7° in width, and these preferences appeared to be independent of each other, when both dimensions were tested on the same cell (42). The dimensional selectivity of DL neurons suggests that DL contributes

to form perception. This hypothesis is consistent with the observation that DL has the most expanded representation of the central visual field (5), where the most acute recognition of form takes place, and the recent discovery that DL is the main source of input to the inferotemporal cortex (58). Inferotemporal cortex has been strongly implicated in the analysis of complex visual stimuli and the learning of visual form discriminations (20, 21). DL is thus an intermediate stage in a major ascending system from layers II and III in V I, to V II, to DL, to IT, and thence to the amygdala and perirhinal cortex (52, 57).

3. Neurons in MT discriminate strongly between stimuli moving in the preferred direction and stimuli moving in the direction 180° opposite, DL neurons generally make this discrimination less well; DM and M neurons generally discriminate this difference in direction of movement very poorly. Most MT neurons resemble the classic directionally selective cells recorded by Barlow et al. (11) in the rabbit retina and observed by many investigators in other species and at other levels in the visual system (23, 50, 56, 60).

4. Neurons in MT respond well to moving texture patterns (arrays of random dots); M neurons respond moderately well; DL and DM neurons respond poorly. All MT cells responded to the moving random dot array and the majority responded better to the array than to the optimally oriented bar moving in its optimal direction and at its optimal velocity. Some neurons in the deeper layers of MT respond well to random dot arrays, but are unresponsive to bar stimuli. All cells tested in M also were responsive to the moving array and one third responded better to the array than to the best bar stimulus. About half the neurons in DL and DM were unresponsive to the moving array and most of the remainder responded poorly to this stimulus. These differences in responsiveness to random dot arrays parallel the discovery, by Hammond and McKay (22) in V I of the cat, that complex cells are responsive to random arrays whereas simple cells are not. It is not known whether there exists a similar dichotomy in responsiveness to textured arrays among neurons in V I in the owl monkey. The striate neurons projecting to MT are the giant cells of Meynert (32, 47, 57), located in the lower parts of layer V, and the neurons located in or near the stria of Gennari. Both these populations of striate neurons are capable of sampling neural activity over relatively large regions. The Meynert cells are by far the largest in striate cortex and their basal dendrites extend for hundreds of micra (13); the layer IVb neurons are located in a dense fiber plexus that was once thought to be the terminations of the geniculate fibers, but has now been shown to be made up to many elements (32). Thus, the striate neurons projecting to

MT could sample over the relatively large regions necessary for the detection of texture arrays. In addition, Montero (37) has demonstrated that there is a relatively large overlap in the projection to MT of different sites in V I by using double-label emulsion autoradiography. MT projects to M (57). MT and M project to the visual pontine nuclei (17, 18), which contain a large proportion of neurons that are especially responsive to moving random dot arrays (9). The pontine nuclei, in turn, project to the cerebellum, where the relayed information presumably contributes to the accomplishment of visuomotor coordination.

5. Neurons in DL, MT, DM and M differ in their distributions of preferred velocities of stimulus movement. The neurons were tested with stimulus velocities ranging from 5°/s to 500°/s. The most common preferred velocity was 10°/s in DL and MT, 50°/s in DM and 100°/s in M. The high preferred stimulus velocities in M may be related to the proportionally large representation of the peripheral visual field in this area (7).

3. Homologous Cortical Visual Areas in Other Species

Beyond V I and V II, the clearest homology is that of MT and similar visual areas in other primate species. The evidence for homology is based on similar location, myeloarchitecture, topography, distinctive anatomical connectivity and visual response properties (1, 2, 8, 10, 12, 21, 46, 49, 53, 54, 57). Owl monkey MT is striate-receptive region of dense myelination coextensive with an orderly map of the visual hemifield (2). A corresponding striate-receptive region of dense myelination coextensive with a similar map of the visual field has been reported in galago (8, 51), marmost (48, 49) and macaque (16, 35, 55, 56). A major source of input to MT in the owl monkey is from V I cells in or near the stria of Gennari and from the giant cells of Meynert located in the lower part of layer V (57). A similar projection occurs in marmoset (47) and macaque (32) from striate cortex cells in the stria of Gennari and the giant Meynert cells. MT is the only known extrastriate cortical target of the Meynert cells (33). Directional selectivity is the principal characteristic of owl monkey MT cell responses, and this has been shown to be true for the ecorresponding region of the macaque (60). The presence of these extensive and detailed similarities in three superfamilies of primates, including primates from both infraorders, indicates that MT probably existed in the early primates ancestral to all living primates. Zeki (61) recently has suggested that MT in the owl monkey is not homologous with the striate-receptive, densely myelinated, directionally selective zone in the posterior bank of the superior temporal sulcus in macaque monkeys; his suggestion is considered in detail in ref. 10.

In our present state of knowledge, it is more difficult to establish clear-cut homologies for the other visual areas found beyond V II in the owl monkey. However, evidence for the homology of several areas is emerging. The principal input to infero-temporal visual cortex in the owl monkey is DL (58). In macaques, a region adjacent to MT is a main input to inferotemporal cortex (14). This region, like DL in the owl monkey, emphasizes the representation of the central visual field (15). MT in both owl monkeys and macaques does not appear to project to inferotemporal cortex. The position of DL between MT and V II in owl monkey is topographically similar to V IV (V 4) in macaques. Neurons in V IV (V 4) have been reported to be specialized for the analysis of color but the percentage of neurons showing color selectivity in V IV (V 4) has ranged from 100% in the original report (59) to less than one third in more recent studies (45, 55). Another recent report suggests that color processing in V IV (V 4) is substantially similar to the color selectivity found in foveal V I and V II (30).

Another potential homology is that of the Ventral Posterior (VP) areas of the owl monkey and the macaque (39, 40). These areas are similar in that they both are long narrow strips that lie immediately anterior to V II on the ventral surface, with this common border corresponding to a representation of the horizontal meridian. In both monkeys, the anterior border of VP corresponds to a discrete band of degeneration following section of the corpus callosum. In both monkeys, the visual field representation in VP appears to be limited to the upper quadrant with the more central portions represented laterally and the more peripheral portions medially. The establishment of potential homolgies for DM and M awaits further investigation.

Outside of primates, it is much more difficult to establish homologies. The last common ancestor of the different mammalian orders lived no more recently than the late Cretaceous period more than 60 million years ago (43). This ancestral mammal had only a very limited development of its neocortex (29). In addition, the adaptive radiation of mammals into different ecological niches with widely divergent behavioral specializations serves to make very difficult the discovery of diagnostic similarities among potentially homologous cortical areas in different mammalian orders. These nonprimate candidates for homology are discussed in reference 10.

4. Significance of Multiple Cortical Areas

Why does the cerebral cortex contain a series of separate representations rather than a single map? In attempting to develop computer analogues of visual perception, Marr elaborated the principle of modular design. Marr (34) stated that any large computation should be broken into a collection of smaller modules as independent as possible from one another. Otherwise, "the process as a whole becomes extremely difficult to debug or improve, whether by a human designer or in the course of natural evolution, because a small change to improve one part has to be accompanied by many simultaneous changes elsewhere." This modular principle has many counterparts in other biological systems. The paleontologist Gregory (19) noted that a common mechanism of evolution is the replication of body parts due to genetic mutation in a single generation, which is followed in subsequent generations by the gradual divergence of structure and functions of the duplicated parts. An analogous idea has been advanced by a number of geneticists. They have theorized that replicated genes escape the pressures of natural selection operating on the original gene and thereby can accumulate mutations, which enable the new gene, through changes in its DNA sequence, to encode for a novel protein capable of assuming new functions (31, 41). Many clearcut examples of gene replication have been discovered (28), and DNA sequence homolgies in replicated genes recently have been established (44). Using this analogy, Allman and Kaas (2, 5) have proposed that the replication of cortical sensory representations has provided the structures upon which new information processing capabilities have developed in the course of evolution. Specifically, it has been argued that existing cortical areas, like genes, can undergo only limited changes and still perform the functions necessary for the animal's survival, but if a mutation occurs that results in the replication of a cortical area, then in subsequent generations the new area can eventually assume new functions through the mechanisms of natural selection, while the original area continues to perform its function.

Acknowledgments

We thank Fran Miezin for developing many of the computer programs used in this study and Leslie Wolcott for drawing the figures. This research was supported by NIH grants NS-00178, NS-12131 and GM-07737, NSF grant BNS-77-15605 and the Pew Memorial Trust.

References

1. ALLMAN, J. M. Evolution of the visual system in the early primates. In: *Progr. Psychobiol. Physiol. Psych.*, vol 7, edited by J. M. Sprague and A. N. Eptsein. New York: Academic Press, 1977, pp. 1–53.
2. ALLMAN, J. M., AND KAAS, J. H. A representation of the visual field in the caudal third of the middle temporal gyrus of the owl monkey *(Aotus trivirgatus)*. *Brain Res.*, 31: 84–105, 1971.
3. ALLMAN, J. M., AND KAAS, J. H. Representation of the visual field in striate and adjoining cortex of the owl monkey *(Aotus trivirgatus. Brain Res.*, 35: 89–106, 1971.
4. ALLMAN, J. M., AND KAAS, J. H. The organization of the second visual area (V II) in the owl monkey: A second order transformation of the visual hemifield. *Brain Res.*, 76: 247–265, 1974.
5. ALLMAN, J. M., AND KAAS, J. H. A crescent-shaped cortical visual area surrounding the middle temporal area (MT) in the owl monkey *(Aotus trivirgatus)*. *Brain Res.*, 199–213, 1974.
6. ALLMAN, J. M., AND KAAS, J. H. The dorsomedial cortical visual area: A third tier area in the occipital lobe of the owl monkey *(Aotus trivirgatus)*. *Brain Res.*, 100: 473–487, 1975.
7. ALLMAN, J. M., AND KAAS, J. H. Representation of the visual field on the medial wall of occiptal-parietal cortex in the owl monkey. *Science*, 191: 572–575, 1976.
8. ALLMAN, J. M., KAAS, J. H., AND LANE, R. H. The middle temporal visual area (MT) in the bushbaby, *Galago senegalensis. Brain Res.*, 57: 197–202, 1973.
9. BAKER, J. F., GIBSON, A., GLICKSTEIN, G., AND STEIN, J. Visual cells in the pontine nuclei of the cat. *J. Physiol., London*, 255: 415–433, 1976.
10. BAKER, J. F., PETERSEN, S. E., NEWSOME, W. T., AND ALLMAN, J. M. Visual response properties of neurons in four extrastriate visual areas of the owl monkey *(Aotus trivirgatus):* A quantitative comparison of the medial (M), dorsomedial (DM), dorsolateral (DL), and middle temporal (MT) areas. *J. Neurophysiol.*, 45: 387–406, 1981.
11. BARLOW, H. B., HILL, R. M., AND LEVICK, W. R. Retinal ganglion cells responding selectively to direction and speed of image motion in the rabbit. *J. Physiol., London*. 173: 377–407, 1964.
12. CAMPBELL, C. B. G., AND HODOS, W. The concept of homology and the evolution of the nervous system. *Brain, Behav. Evol.*, 3: 353–367, 1970.
13. CHAN-PALAY, V., PALAY, S. L., AND BILLINGS-GAGLIARDI, S. M. Meynert cells in the primate visual cortex. *J. Neurocytol.*, 3: 631–658, 1974.

14. DESIMONE, R., FLEMING, J., AND GROSS, C. G. Prestriate afferents to inferior temporal cortex: an HRP study. *Brain Res.*, in press, 1981.

15. DESIMONE, R. AND GROSS, C. G. Visual areas in the temporal cortex of the macaque. *Brain Res.*, 178: 363–380, 1979.

16. GATTASS, R., AND GROSS, C. G. A visuotopically organized area in the posterior superior temporal sulcus of the macaque. *ARVO Annual Meeting Abstr.*, 1979, p. 184.

17. GLICKSTEIN, M., COHEN, J., DIXON, B., GIBSON, A., HOLLINS, M., LA BOSSIERE, E., AND ROBINSON, F. Corticopontine visual projection in the macaque monkey. *J. Comp. Neurol.*, 190: 209–230, 1980.

18. GRAHAM, J., LIN, C.-S., AND KAAS, J. H. Subcortical projections of six visual cortical areas in the owl monkey, *Aotus trivirgatus*. *J. Comp. Neurol.*, 187: 557–580, 1979.

19. GREGORY, W. K. Reduplication in evolution. *Quart. Rev. Biol.*, 10, 272–290, 1935.

20. GROSS, C. G. Visual functions of inferotemporal cortex. In: *Handbook of Sensory Physiology, VII/3 B* edited by R. Jung. Berlin: Springer, 1973, p. 451–482.

21. GROSS, C. G., BRUCE, C. J., DESIMONE, R., FLEMING, J., AND GATTASS, R. Three visual areas of the temporal lobe. *This volume,* chapter 8.

22. HAMMOND, P., AND MACKAY, D. M. Differential responsiveness of simple and complex cells in cat striate cortex to visual texture. *Exptl. Brain Res.*, 30: 106–154, 1977.

23. HUBEL, D. H., AND WIESEL, T. N. Receptive fields, binocular interaction and functional architecture in the cat's visual cortex. *J. Physiol., London,* 60: 106–154, 1962.

24. HUBEL, D. H., AND WIESEL, T. N. Receptive fields and functional architecture in two non-striate visual areas (18 and 19) of the cat. *J. Neurophysiol.*, 28: 229–289, 1965.

25. HUBEL, D. H., AND WIESEL, T. N. Visual area of the lateral suprasylvian gyrus (Clare-Bishop area) of the cat. *J. Physiol., London,* 202: 251–260, 1969.

26. HUBEL, D. H., AND WIESEL, T. N. Functional architecture of macaque monkey visual cortex. *Proc. Roy. Soc., London, B,* 198: 1–59, 1977.

27. IMIG, T. J., RUGGERO, M. A., KITZES, L. M., JAVEL, E., AND BRUGGE, J. F. Organization of auditory cortex in the owl monkey *(Aotus trivirgatus). J. Comp. Neurol.*, 171: 111–128, 1977.

28. INGRAM, V. M. *The Hemoglobins in Genetics and Evolution.* New York: Columbia Univ. Press, 1963.

29. JERISON, H. *Evolution of the Brain and Intelligence.* New York: Academic Press, 1973.

30. KRUGER J., AND GOURAS, P. Spectral selectivity of cells and its dependence on slit length in monkey visual cortex. *J. Neurophysiol.*, 43: 1055–1069, 1980.

31. LEWIS, E. B. Pseudoallelism and gene evolution. *Cold Spring Harbor Symp. Quant. Biol.*, 16: 159–174, 1951.

32. LUND, J. S., LUND, R. D., HENDRICKSON, A. E., BUNT, A. H., AND FUCHS, A. F. The origin of efferent pathways from the primary visual cortex, area 17, of the macaque monkey as shown by retrograde transport of horseradish peroxidase. *J. Comp. Neurol.*, 164: 287.304, 1976.

33. LUND, J. S., HENRY, T. H., MACQUEEN, C. L., AND HARVEY, A. R. Anatomical organization of the primary visual cortex (area 17) of the cat. A comparison with area 17 of the macaque monkey. *J. Comp. Neurol.*, 184: 599–618, 1979.

34. MARR, D. Early processing of visual information. *Phil. Trans. Roy. Soc., London, Series B*, 275: 484–519, 1976.

35. MAUNSELL, J. H. R., BIXBY, J. L., AND VAN ESSEN, D. C. The middle temporal (MT) area in the macaque: Architecture, functional properties and topographic organization. *Soc. Neurosci. Abstr.*, 5: 796, 1979.

36. MERZENICH, M. M., KAAS, J. H., SUR, M., AND LIN, C.-S. Double representation of the body surface within cytoarchitectonic areas 3b and 1 in SI in the owl monkey *(Aotus trivirgatus)*. *J. Comp. Neurol.*, 181: 41–74, 1978.

37. MONTERO, V. M. Patterns of connections from the striate cortex to cortical visual areas in superior temporal sulcus of macaque and middle temporal gyrus of owl monkey. *J. Comp. Neurol.*, 189: 45–55, 1980.

38. MYERSON, J., MANIS, P. B., MIEZIN, F. M., AND ALLMAN, J. M. Magnification in striate cortex and retinal ganglion cell layer of owl monkeys: A quantitative comparison. *Science*, 198: 855–857, 1977.

39. NEWSOME, W. T., AND ALLMAN, J. M. The interhemispheric connections of visual cortex in the owl monkey, *Aotus trivirgatus*, and the bushbaby, *Galago senegalensis. J. Comp. Neurol.*, 194: 209–233, 1980.

40. NEWSOME, T. W., MAUNSELL, J. H. R., AND VAN ESSEN, D. C. Areal boundaries and topographic organization of the ventral posterior area (VP) of the macaque monkey. *Soc. Neurosci. Abstr.*, 6: 579, 1980.

41. OHNO, S. *Evolution by Gene Duplication.* New York: Spring, 1970, p. 1–60.

42. PETERSEN, S. E., BAKER, J. F., AND ALLMAN, J. M. Dimensional selectivity of neurons in the dorsolateral visual area of the owl monkey. *Brain Res.*, 197: 507–511, 1980.

43. ROMER, A. S. *Vertebrate Paleontology.* Chicago: Univ. of Chicago Press, 1966, pp. 1–460.

44. ROYAL, A., GARAPIN, A., CAMIL, B., PERRIN, F., MANDEL, J. L., LEMEUR, J., BREGEGEGRE, F., LEPENNEC, J. P., CHAMBON, P., AND KOURILSKY, P. The ovalbumin gene region: Common features in the organization of three genes expressed in chicken oviduct under hormonal control., *Nature*, 279: 125–132, 1979.

45. SCHEIN, S. J., MARRACCO, R. T., AND DE MONASTERIO, F. M. Spectral properties of cells in the prestriate cortex of monkey. *Soc. Neurosci. Abstr.*, 6: 580, 1980.

46. SIMPSON, G. G. *Principles of Animal Taxonomy.* New York: Columbia Univ. Press, 1961.

47. SPATZ, W. B. An efferent connection of the solitary cells of Meynert. A study with horseradish peroxidase in the marmoset, *Callithrix. Brain Res.,* 92: 450–455, 1975.

48. SPATZ, W. B. Topographically organized reciprocal conections between areas 17 and MT (visual area of the superior temporal sulcus) in the marmoset *(Callithrix jacchus). Exptl. Brain Res.,* 27: 559–572, 1977.

49. SPATZ, W. B., AND TIGGES, J. Experimental-anatomical studies on the Middle Temporal Visual Area (MT) in primates. I. Efferent cortico-cortical connections in the marmoset *(Callithrix jacchus). J. Comp. Neurol.,* 146: 451–563, 1972.

50. SPEAR, P. D., AND BAUMANN, T. P. Receptive-field characteristics of single neurons in lateral suprasylvian visual area of the cat. *J. Neurophysiol.,* 38: 1403–1420, 1975.

51. TIGGES, J., TIGGES, M., AND KALAHA, C. S. Efferent connections of area 17 in *Galago. Amer. J. Phys. Anthropol.,* 38: 393–398, 1973.

52. TURNER, B. H., MISHKIN, M., AND KNAPP, M. Organization of amygdalopetal projections from modality-specific cortical association areas in the monkey. *J. Comp. Neurol.,* 193: 147–184, 1980.

53. UNGERLEIDER, L. G., AND MISHKIN, M. The striate projection zone in the superior temporal sulcus of *Macaca mulatta:* Location and topographic organization. *J. Comp. Neurol.,* 188: 347–366, 1979.

54. VAN ESSEN, D. C. Visual cortical areas. In: *Ann. Rev. Neurosci.,* Vol. 2, edited by W. M. Cowan. Palo Alto: Annual Reviews, Inc., 1979, p.

55. VAN ESSEN, D. C., AND ZEKI, S. M. The topographic organization of rhesus monkey prestriate cortex. *J. Physiol., London,* 277: 193–226, 1978.

56. WELLER, R. E., KAAS, J. H. Connections of striate cortex with the posterior bank of the superior temporal sulcus in macaque monkeys. *Soc. Neurosci. Abstr.,* 4: 650, 1978.

57. WELLER, R. E., AND KAAS, J. H. Cortical and subcortical connections of visual cortex in primates. *This volume,* chapter 13.

58. WELLER, R. E., AND KAAS, J. H. Connections of the dorsolateral visual area (DL) of extrastriate visual cortex of the owl monkey *(Aotus trivirgatus). Soc. Neurosci. Abstr.,* 6: 579, 1980.

59. ZEKI, S. M. Colour coding in the superior temporal sulcus of rhesus monkey visual cortex. *Brain Res.,* 422–427, 1973.

60. ZEKI, S. M. Functional organization of visual area in the posterior bank of the superior temporal sulcus of the rhesus monkey. *J. Physiol., London,* 236: 549–573, 1974.

61. ZEKI, S. M. The response properties of cells in the middle temporal area (area MT) of owl monkey visual cortex. *Proc Roy. Soc. London, B.* 207: 239–248, 1980.

Chapter 8

Cortical Visual Areas of the Temporal Lobe

Three Areas in the Macaque

C. G. Gross, C. J. Bruce, R. Desimone, J. Fleming and R. Gattass

Department of Psychology, Princeton University, Princeton, New Jersey

1. Introduction

Primates are very visual animals. Thus, it is not surprising that over half of their cerebral cortex is devoted primarily to visual function (67). In this essay, we will consider three rather different cortical visual areas, all of which lie within the temporal lobe of the macaque, viz., the Middle Temporal Area (or MT), the Inferior Temporal Area (or IT) and the Superior Temporal Polysensory Area (or STP). Their locations are shown in Fig. 8.1.

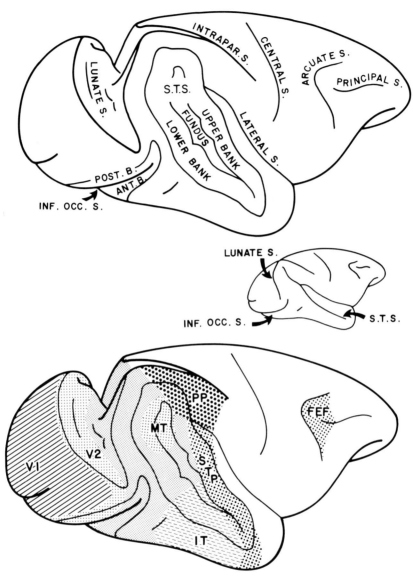

Fig. 8.1. The small drawing shows a lateral view of the cerebral hemisphere of the macaque. The larger drawings are traced from a photograph of a brain, in which the superior temporal sulcus (STS), the lunate sulcus and the inferior occipital sulcus (INF. OCC. S.) were opened up to show their banks and fundi. The lower drawing indicates the location of the areas discussed in this paper. The area shown in grey is additional visually responsive prestriate cortex. ANT. B., anterior bank; IT, inferior temporal cortex; MT, middle temporal area; POST. B., posterior bank; PP, posterior parietal cortex; STP, superior temporal polysensory area.

For each area we will consider (a) its location and architectonic characteristics, (b) the visual properties of its neurons, (c) its afferents and efferents and (d) the behavioral effects of its removal. Then we will suggest that the various extrastriate cortical visual areas may be divided into three functional classes and that each of the three temporal areas discussed are representative of one of these classes.

2. The Middle Temporal Area (MT)

2.1. Location and Architectonics

MT lies in the floor and adjacent portion of the lower bank of the caudal third of the superior temporal sulcus as shown in Figs. 8.2 and 8.3 (22, 23, 46, 65, 67). It falls within cytoarchitectonic area OA of von Bonin and Bailey (68) and area 19 of Brodmann (10); thus, it is a portion of prestriate cortex. In fiber-stained sections, it is readily distinquishable from the adjacent cortex by heavy myelination in layers four through six.

Although this area is located at the junction of the occipital, temporal and parietal lobes of the macaque, it has been called the Middle Temporal area (MT) (1, 46, 66, 67, 70) because on several grounds it appears homologous to an area designated MT in other primates (1, 2, 3, 63, 66). In these species, MT is indeed in the middle of the temporal lobe. In the macaque, this area has also been called the "STS-movement area" (77, 81, 82), and the "striate-projection zone in the superior temporal sulcus" (23, 65).

2.2. Neuronal Properties

MT neurons respond only to visual stimuli. Their receptive fields are organized to provide a topographic representation of the contralateral visual hemifield (22, 23, 46, 65, 67). The visuotopic organization is similar to that of striate cortex, that is, a first-order transformation of the visual field. The representation of the vertical meridian partially surrounds the area and lies near the bottom of the lower bank of the superior temporal sulcus (STS). The representation of the horizontal meridian runs across the floor of STS, the upper visual field being anterior and the lower visual field posterior (see Figs. 8.2 and 8.3). As in striate cortex, the representation of the central visual field is greatly magnified. Receptive field size increases strikingly with eccentricity so that receptive fields with

A

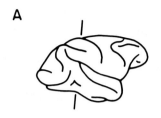

B

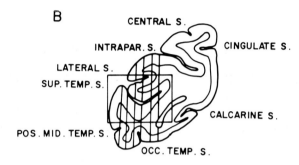

C

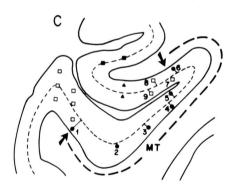

● VISUAL RESPONSES WITHIN MT
□ VISUAL RESPONSES OUTSIDE MT
▲ POLYSENSORY RESPONSES
■ AUDITORY RESPONSES

FIG.   8.2. Progression of receptive fields in MT. A, lateral view of hemisphere showing level of cross section. B, cross-section showing the location of the electrode penetrations and the portion of the superior temporal sulcus (STS) enlarged below. C, section through the superior temporal sulcus showing the location of MT (heavily dashed line between arrows) and of recording sites within MT (1–6) and in the surrounding areas. The fourth cortical layer (thinner dashed line) is also shown and the projection of the recording sites onto it. D, location of the receptive field centers re-

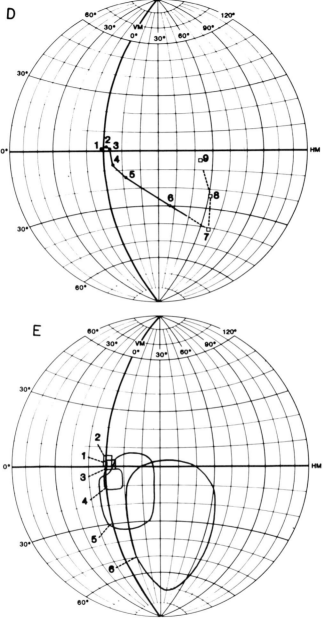

FIG. 8.2. (*continued*)
corded at sites 1–9. Note that, as one moves from site 1 to site 6, the receptive field centers move systematically from the center of gaze into the periphery. Between site 6 and 7 the myeloarchitecture changes; sites 7–9 are outside of MT, and from site 7 to site 9 the field centers move towards the horizontal meridian. E, receptive fields recorded at sites 1–6. Note the rapid increase in receptive field size with eccentricity (22).

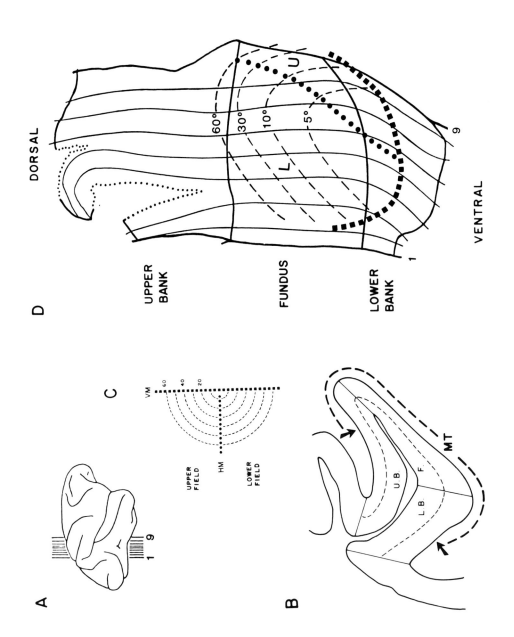

FIG. 8.3. Visuotopic organization of MT. A, lateral view of hemisphere showing location of sections used to construct an unfolded map of the portion of the STS that includes MT. Wires were bent to conform to layer 4 of each section and then were attached with scaled cross pieces to form a three-dimensional model of the relevant region of STS. The model was then flattened by cutting the minimum number of cross pieces to form a two-dimensional surface. B, section 7 showing MT (heavy dashed line) and location of upper bank (UB), fundus (F) and lower bank (LB) of STS. (This section was shown in Fig. 8.2.) C, map of visual hemifield showing meridians and isoeccentricity lines. D, tracing of flattened model constructed from the sections. The locations of the receptive fields were plotted on the tracing. On the basis of the coordinates of the receptive field centers, the vertical meridian (large dots), the horizontal meridian (squares) and the isoeccentricity lines (dashed) were drawn. The dotted lines show the discontinuities produced by flattening the three dimensional model. The numbered lines (1–9) correspond to the coronal sections shown in A (22).

their centers 10–20° from the fovea are often so large that their medial borders are on or near the vertical meridian (Fig. 8.2E). Perhaps because of the larger receptive fields of its neurons and its relatively small total area, the visuotopic organization of MT is somewhat cruder than that of either V I (V I), or V2 (V II) (22, 23).

Unlike neurons in other prestriate areas, a large proportion of MT neurons are sensitive to the direction of movement of a stimulus, but are relatively insensitive to form, orientation, size or color (46, 77, 80, 81, 82). Some MT neurons apparently respond to changing disparity or to stimuli with expanding or contracting contours (78); these units may signal movement in depth. Finally, MT neurons, as in other prestriate areas, generally respond equally to stimulation of either eye (81).

2.3. Anatomical Connections

MT receives a topograhically organized projection from striate cortex (65, 70); thus, it is a "striate-recipient" area, as is V2 (V II) (74). MT also receives projections from V2 (V II) (79) and from the anterior parts of the inferior and lateral pulvinar (8). Since these portions of the pulvinar apparently receive a tectal projection (7, 8), MT may be the site of converging inputs from the geniculostriate and tectofugal systems.

MT projects back to striate cortex in a topographically organized fashion (70). Otherwise the targets of MT axons have not been investigated in the macaque.

2.4. Behavioral Effects of Removal

Since the location of MT in the macaque has only recently been described, there are very few data on the effects of its selective removal. MT lesions do not seem to impair acquisition or retention of visual pattern discrimination problems (4, 64). However, they do impair a monkey's ability to pick a small food pellet out of a narrow slot and to detect and grasp a loose peanut mounted on a background of fixed peanuts (4). These results suggest that MT, unlike IT (see next section), may not be directly involved in visual pattern learning, but rather in other visual functions, perhaps of a visuomotor or strictly sensory nature. A similar suggestion has been made by Wilson and her colleagues (72).

3. Inferior Temporal Cortex (IT)

3.1. *Location and Architectonics*

IT lies on the inferior convexity of the temporal lobe. It extends from about 6 mm anterior to the ascending portion of the inferior occipital sulcus to within a few millimeters of the temporal pole and from the floor of the superior temporal sulcus to the depths of the occipitotemporal sulcus (21). IT corresponds closely to cytoarchitectonic area TE, as defined by von Bonin and Bailey and others (21, 62, 68).

3.2. *Neuronal Properties*

IT neurons respond only to visual stimuli. Unlike V1 (V I), V2 (V II) and MT, inferior temporal cortex is not visuotopically organized (21, 27, 33). Rather, throughout IT, receptive fields almost always include the center of gaze and in a majority of cases extend well across the vertical meridian into both visual half-fields (Fig. 8.4). IT receptive fields are particularly large for receptive fields that include the center of gaze; the median receptive field size is approximately $25 \times 25°$ (21,27,29,33). Unlike neurons in striate cortex, most IT neurons respond better to complex objects than to slits of light or edges (18, 33). The majority appear selective for features of objects such as their shape, texture or color. Responses of one such neuron are shown in Fig. 8.5. A few IT units appear selective for specific objects such as hands or faces. The stimulus selectivity of a given IT neuron is similar throughout its large receptive field (32, 33). That is, they show stimulus equivalence across retinal translation. These properties of IT neurons indicate that they play a role in analyzing the global aspects of a complex object such as its shape, rather than in analyzing local features such as the retinal locus of edges and borders.

Within IT, there are two regions that have larger receptive fields and are cytoarchitectonically distinguishable from the rest of IT. The first is in the most rostral part and the second lies along the dorsal border of IT in STS (21).

3.3. *Anatomical Connections*

The visual properties of IT units depend on a multisynaptic corticocortical input from striate cortex (58). Striate cortex does not pro-

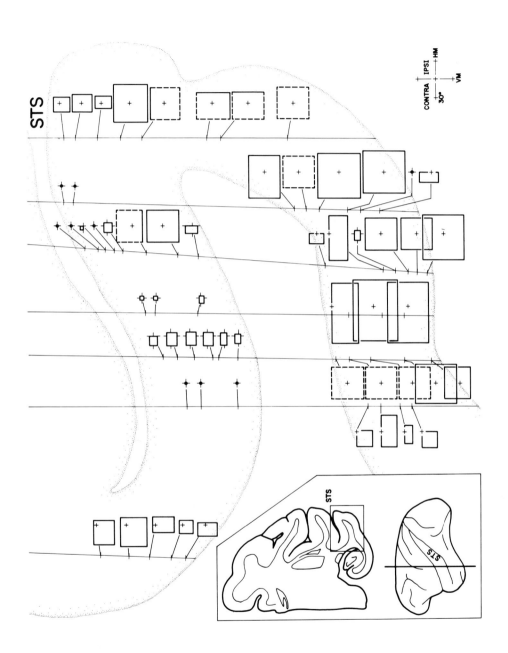

FIG. 8.4. Receptive fields of single neurons on a series of penetrations through inferior temporal cortex. Receptive fields are represented by best-fitting rectangles. In some cases, receptive field borders extending beyond the 60 × 60° tangent screen were not determined and are represented by dashed lines; receptive fields smaller than 8° were all located at the center of gaze and are shown as dots. The meridians of the visual field are represented by vertical and horizontal lines. Note that all the receptive fields include the center of gaze and that many fields extend far into the ipsilateral hemifield. Although neurons with similar receptive fields tend to cluster, there is no sign of any visuotopic organization. The inset shows the location of the section on a lateral view of the brain and the portion of the section enlarged in the main figure. The scale applies to the receptive field dimensions.(21)

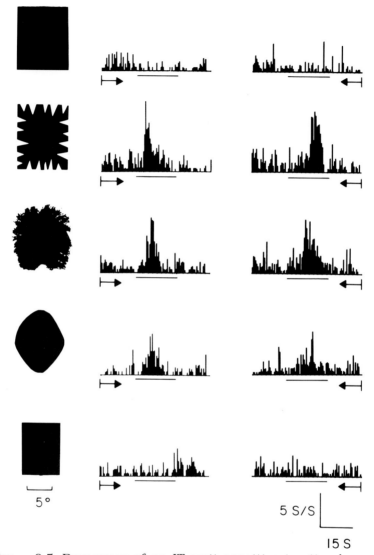

FIG. 8.5 Responses of an IT unit sensitive to stimulus contour. When tested with a set of complex three-dimensional objects, this unit responded maximally to objects with irregular edges. Responses shown here are to two-dimensional, high-contrast, dark patterns. The stimuli were moved along the horizontal meridian at 4.5°/s and appeared within a 15° "window" centered on the fovea. The line under each histogram represents the period in which at least one edge of the largest stimulus was visible through the window. Each of the histograms is based on five interleaved trials. Orientation and direction of movement were not relevant (18).

ject directly to IT cortex (16, 20, 41, 74). Rather, striate cortex projects to V2 (V II) and other striate-recipient prestriate areas. These project to more anterior prestriate areas (including Zeki's V4 complex) (20, 75). Finally, these anterior prestriate areas project directly to IT cortex (Fig. 8.6) (20). Thus, there are at least two processing stages between striate and IT cortex. In addition to this ipsilateral route, IT receives visual information from contralateral anterior prestriate cortex and the contralateral IT cortex by way of both the splenium and the anterior commissure (20, 30, 52, 53, 76). IT cortex also receives projections from visually responsive areas in lateral frontal cortex (40) and the posterior portion of the lateral pulvinar (8).

IT cortex projects to a variety of cortical and subcortical sites. IT projects to lateral frontal cortex, anterior prestriate cortex and the posterior part of the lateral pulvinar (40, 41, 71). IT cortex also projects to the STS polysensory area and parahippocampal region (40, 41) and to several subcortical sites including the amgydala, the tail of the caudate nucleus, the dorsomedial nucleus of the thalamus and the deep layers of the superior colliculus (71).

3.4. Behavioral Effects of Removal

Bilateral removal of inferior temporal cortex produces a severe deficit in visual discrimination learning. This deficit occurs in the absence of any change in sensory status; visual acuity, a variety of other psychophysical thresholds and the integrity of the visual fields all remain normal after IT lesions, as does discrimination learning in other modalities than the visual (15, 26, 27).

The discrimination impairment after IT lesions is both in postoperative acquisition and postoperative retention of visual discrimination tasks. It occurs with a great variety of discriminanda, such as stimuli differing in pattern, shape, size, brightness and color and with a variety of training methods. There are several conditions under which the impairment may be reduced or absent: when the task is very simple; when the discriminanda are patterns differing in orientation by 90 or 180°; when the animals receive extensive preoperative training and when the lesions are made in infancy (26, 28).

Different partial lesions of IT cortex produce somewhat different deficits. The deficits after posterior IT lesions have been characterized as perceptual in nature, whereas the more anterior ones seem to be mnemonic (14, 27, 45, 48).

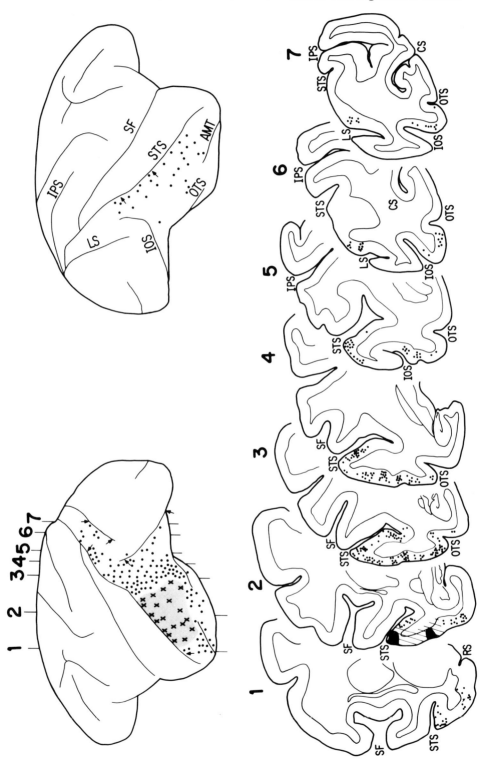

Fig. 8.6. Distribution of labeled cells in occipital and temporal cortex following injections of the retrograde tracer horseradish peroxidase (HRP) into inferior temporal cortex. In the lateral reconstructions, × indicates the sites of the iontophoretic injections and the shaded area indicates the area of spread of HRP. The relative density of labeled cells is indicated by dots and arrows delimit where labeled cells were found in a bank of a major sulcus. Labeled cells were also found in both banks of the anterior middle temporal sulcus and in the occipitotemporal sulcus, mainly in the lateral bank. In the coronal sections, the black areas indicate the sites of the injections, the hatched areas indicate the spread of HRP and each dot represents an individual labeled cell. Note that HRP-labeled cells were found throughout IT itself (outside the injection area), but were not found in the polysensory areas that surround IT dorsally, anteriorly and ventrally. Posterior to IT, labeled cells were found in the anterior parts of prestriate cortex. This prestriate region that projects to IT is mainly devoted to the central visual field and does not include any of the prestriate areas that receive direct projections from striate cortex, such as MT or V2 (20) .

4. Superior Temporal Polysensory Area (STP)

4.1. Location and Architectonics

The area we have termed the superior temporal polysensory area lies chiefly in the upper bank of the middle and rostral portions of the superior temporal sulcus immediately dorsal to IT (11, 12, 21). Anteriorly it crosses the floor of STS and extends onto the lateral surface at the temporal pole. It corresponds approximately to Area T3 defined by Jones and Burton (39) on grounds of architectonics and pulvinar afferents.

4.2. Neuronal Properties

Virtually all the neurons in STP are responsive to visual stimuli. However, in contrast to both MT and IT, almost half the neurons in STP are also responsive to auditory or somesthetic stimuli and many respond to stimulation in all three modalities (Figs. 8.7 and

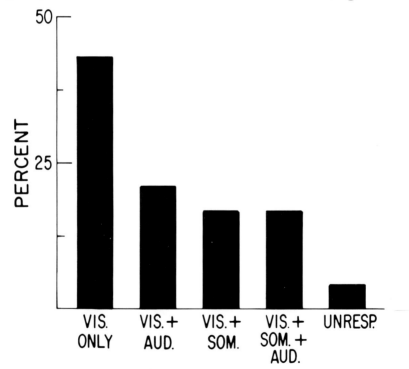

FIG. 8.7. Frequency of STP neurons responsive to different sensory modalities. The percentages are based on 382 neurons tested for visual (VIS), auditory (AUD), and somesthetic (SOM) responses. UNRESP, unresponsive to visual, auditory, and somesthetic stimuli (11).

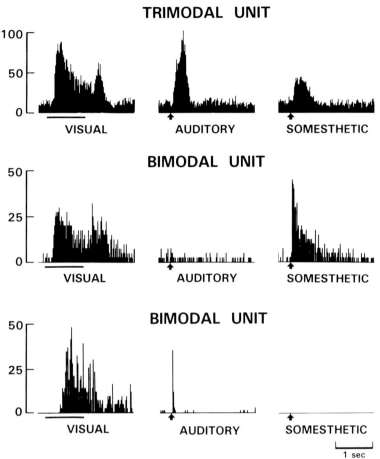

FIG. 8.8. Poststimulus time histograms showing the responses of three STP neurons to a visual stimulus, to a click and to a mechanical tap on the bottom of the foot. The visual stimulus for the upper and middle histograms was a vertical slit of light and for the bottom histogram a color slide of a monkey's face. The vertical scale in this and subsequent histograms represents the number of impulses per second. The horizontal lines indicate the presentation of the visual stimuli and the arrows the presentation of the auditory and somesthetic ones (11).

8.8). It was this multimodal responsivity that led us to name the area the "superior temporal polysensory area" (11, 12).

The visual receptive fields in STP are extraordinarily large, even when compared with those in inferior temporal cortex (Fig. 8.9). The majority of them extends into the monocular crescent of both eyes; the median value for the contralateral extent is 80°, for the ipsilateral extent 70°, for the upperward extent 50° and for the downward extent 55°. Thus, for most STP units the receptive field approaches the size of the monkey's visual field. Obviously, this

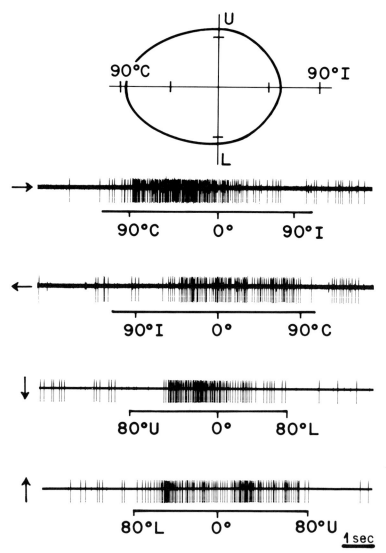

Fig.   8.9. Typical visual receptive field of a STP neuron (upper) and responses (lower) to a stimulus moved along each of the meridians in the direction indicated by the arrows. In the receptive field plot, the horizontal and vertical lines represent the meridians. The stimulus was a 5° white square moving at approximately 35°/s along a circular path around the animal's head. The scale under each trace indicates the location of the stimulus. Only the contralateral eye was stimulated. In each case the stimulus began moving outside the visual field, moved through the receptive field and then left the visual field. C, contralateral; I, ipsilateral; L, lower; U, upper (11).

precludes any type of visuotopic organization. Responses are usually similar throughout the receptive field and are not more vigorous at the center of gaze, as is the case for IT neurons.

In contrast to IT units (but more similar to MT units), the majority of STP units is not sensitive to the size, shape, orientation, color or contrast of the visual stimulus. STP units respond particularly well to moving stimuli and about half show one of three types of direction selectivity. One class of units has conventional or uniform direction preferences, such as leftward or downward throughout the receptive field. Another class of units has an unusual, nonuniform type of direction preference. About half of the units in this class respond to any movement directed away from the fovea (Fig. 8.10). The other half prefers stimulus movement towards the fovea. Finally, a third class responds to stimulus motion in depth: to movement either toward or away from the animal (Fig. 8.11).

A minority of STP units appears particularly sensitive to monkey and human faces (Fig. 8.12). Similar units have been described in both inferior temporal (18, 33) and lateral frontal cortex (56).

4.3. Anatomical Connections

STP receives afferents from inferior temporal cortex, superior temporal auditory association cortex and posterior parietal cortex (38, 40, 41, 62). These connections may be sources of its visual, auditory and somesthetic sensitivity, respectively. It also receives afferents from a variety of other cortical and subcortical areas, including cingulate cortex, the parahippocampal region, the frontal eye fields, the amygdala and the medial pulvinar (38, 39). The connections with both the frontal eye fields and posterior parietal cortex are reciprocal ones (17,47).

As mentioned previously, the visual responses of IT neurons are dependent on the corticocortical input they receive from striate cortex (58). This is apparently also true of V2 (V II) neurons (60). Surprisingy, this is not true of STP neurons (19). Following unilateral striate lesions, over two-thirds of STP neurons remain visually responsive in the "cortically blind" half-field. However, in these units, response strength, receptive field size and direction specificity are all reduced in the half-field contralateral to the lesion. Preliminary evidence indicates that the addition of a superior colliculus lesion to the striate cortex lesion eliminates this visual responsiveness. Thus, apparently both striate cortex and the superior colliculus contribute to the visual properties of STP neurons. After total bilateral removal of striate cortex, macaques retain or recover an impressive degree of visually guided behavior (e.g., 37, 50,

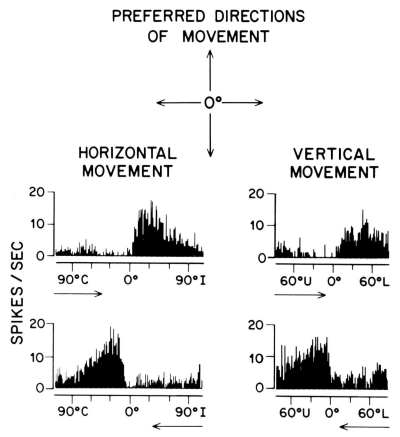

FIG. 8.10. Responses of a STP unit selective for stimulus movement away from the center of gaze. The histograms show the responses to a 5° × 3° stimulus moving on a circular path around the animal's head along the horizontal meridan at 22°/s or the vertical meridian at 15°/s. The arrows indicate the direction of stimulus movement. Only the contralateral eye was stimulated (11).

69). It is possible that the visual functions of STP that remain after striate lesions contribute to this residual vision.

4.4. Behavioral Effects of Removal

In one of the few studies of lesions largely confined to STP, no deficits were found in visual discrimination learning (55). However, these animals did have a severe deficit in a visuospatial task requiring them to "unstring" a piece of food from a wire with multiple bends (55) and in a multimodal task involving a visual–auditory association (54).

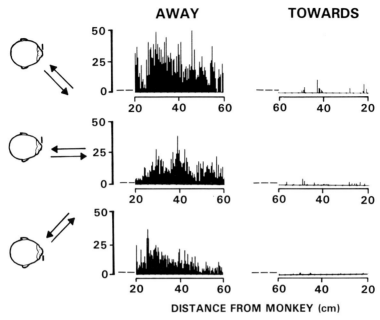

Fig. 8.11. Responses of a STP unit selective for stimulus movement away from the animal. The stimulus was a color photograph of a monkey face moving at approximately 20 cm/s along the paths indicated by the arrows. The dashed lines indicate the mean spontaneous rate in the interstimulus intervals. Note that movement toward the animal reduces the rate of firing. One eye was occluded as shown on the left. The receptive field extended approximately 65° into each visual hemifield. The ordinate is in spikes/s (11).

5. An Hypothesis: Three Classes of Extrastriate Visual Areas

The middle temporal area (MT), inferior temporal cortex (IT) and the superior temporal polysensory area (STP) represent only a minority of the extrastriate visual areas devoted primarily or exclusively to visually guided behavior. For most of the extrastriate visual areas, in contrast to striate cortex, we do not have a complete description of their inputs and outputs, we know little about the behavioral effects of their selective removal and their neuronal properties have only been examined under a restricted set of stimulus and behavioral conditions. Even at this early stage, however, it may be helpful to try to group them into functional classes. The extrastriate visual areas appear to fall into three classes and, it will be argued, each of the temporal areas discussed in this essay exemplifies one of these classes.

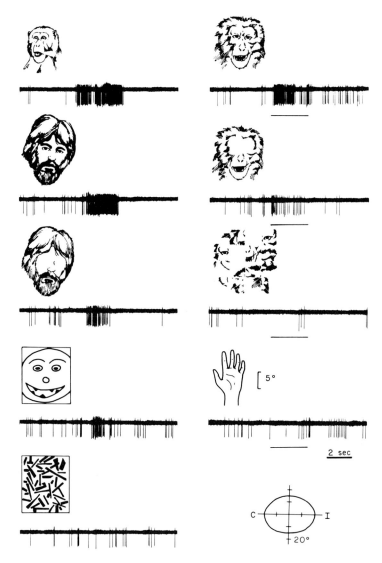

Fig.   8.12. Responses of an STP unit that responded better to pictures of faces than to a variety of other stimuli. Removing the eyes or representing the face as a caricature reduced the response. Cutting the picture into sixteen pieces and re-arranging the pieces eliminated the response. The stimuli represented on the left were traced from a color photograph (monkey face), black and white photographs (human face) or drawings (caricature and random bars) which were swept across the fovea at 10°/s. The stimuli represented on the right were traced from color slides which were projected centered on the fovea for 3 s, indicated by the horizontal bars. All of the unit records are representative ones chosen from a larger number of trials. The receptive field is illustrated on the lower right (11).

The first class consists of the various prestriate areas within cytoarchitectonic areas OB and OA. This class may be defined as being exclusively visual and visuotopically organized, i.e., its receptive fields are arranged to form a spatial representation of the contralateral visual field. Although varying in their visual topography, the members of this class are characterized, like striate cortex, by a magnification of the central portion of the visual field. They receive either direct input from striate cortex or direct input from a striate-recipient area (66, 81, 82). Preliminary neurophysiological evidence indicates that neurons in some of these areas may be relatively specialized for the analysis of such stimulus features as direction of movement and size (1, 36, 81, 82). MT is an example of this first class as are V2 (V II) and V3 (V III).

The second class consists of inferior temporal cortex. This class may be defined as being exclusively visual and *not* visuotopically organized (21, 31, 33); the receptive fields almost always include the center of gaze and usually extend well into both visual half-fields. This class receives input from nonstriate-recipient members of the first class. Inferior temporal neurons are usually sensitive to several stimulus dimensions and often have highly complex trigger features. Although this class consists only of inferior temporal cortex, it appears subdividable in terms of its cytoarchitectonics, its connections, its neuronal properties and the behavioral effects of partial lesions (17, 21, 26, 45).

The third class may be defined as being not exclusively visual and having little or no visuotopic organization. It is characterized by having neurons that usually (a) have very large visual receptive fields, (b) are not sensitive to form or color *per se* and (c) are sensitive to auditory or somesthetic stimulation or are related to motor acts or both. This class includes the superior temporal polysensory area described above, the frontal eye fields (area 8), posterior parietal cortex (area 7), and perhaps the parahippocampal region (12, 21, 49, 51, 57). Furthermore, the members of this class are reciprocally connected with each other.

Although all three classes of visual areas receive projections from both the geniculostriate and tectofugal visual pathways, only the geniculostriate system appears crucial for the visual properties of the first two classes (19, 25, 42, 60). By contrast, the superior colliculus appears intimately involved in the functions of the third class.

What may be the functions of these three classes of visual areas?

Removal of inferior temporal cortex, the second class, in both man and monkey produces impairments in visual perception and visual memory in the absence of any visuosensory changes (26, 27).

These results indicate that the second class is involved in pattern perception and recognition. The fact that IT neurons are particularly concerned with the center of gaze, have complex trigger features and show stimulus equivalence across retinal translation (18, 32, 61) further supports this view.

Several lines of evidence suggest that the third class of visual areas is particularly involved in relating visual stimuli to movement, that is, in spatial orientation and visuomotor (especially oculomotor) mechanisms. Posterior parietal lesions impair spatial orientation, visuoconstructive skills, visually guided locomotion and reaching and they produce various neglect and inattention symptoms; posterior parietal neuronal activity is correlated with both eye movements and orienting to and reaching toward visual stimuli (34, 35, 42, 51, 57). Frontal eye field lesions produce both visual neglect and deficits on spatial tasks and its neurons discharge in relation to eye movements (9, 43, 44, 59, 73). STP lesions produce both a neglect syndrome and visuomotor disturbances; the very large receptive fields of its neurons and their unusual directional selectivities suggest a role in spatial orientaton (11, 12, 54, 55).These three "third class" areas have two other properties that further suggest that they form a system for linking sensory input with action. The first is that they are closely associated with the superior colliculus and the superior colliculus is a sensory-motor structure concerned with oculomotor, orientation and localization functions. The second is that, like the superior colliculus, neurons in all three areas respond to nonvisual as well as visual stimuli (12, 49, 51). This may reflect the fact that orientation and localization, although especially visual in primates, are actually supramodal functions.

Areas of the first class, the prestriate areas, are at an earlier stage of visual processing than the other two classes. They appear concerned with the analysis of specific visual features at a specific retinal locus (13, 81). Large lesions of prestriate cortex that include the foveal representations of several prestriate areas produce visual learning deficits even more severe than those following IT lesions (27). The converging outputs of the various prestriate areas may provide the basis upon which the second class elaborates its pattern recognition functions and the third class its visuomotor functions.

Finally, it should be noted that we have been concerned exclusively with cortical visual mechanisms. However, there are subcortical mechanisms also concerned with stimulus analysis and visuomotor function. For example, the pulvinar projects to all three classes of cortical visual areas and itself contains several visuotopically organized areas and several classes of stimulus feature analyzers (5, 6, 8, 24). Certainly, an understanding of the mechanisms of visually guided behavior will require not only understanding the cortical visual areas but also their interaction with subcortical visual structures.

Acknowledgments

We would like to thank A. P. B. Sousa, V. Ingalls and G. Barber for their critical comments, K. Aragão de Carvalho for help with the figures and B. Pinkham for typing.

Preparation of this chapter was supported by NIH Grant MH-19420 and NSF Grant BNS 79-05589. CB was supported by NIH Grant NS-05804, RD by NSF Grant SPI-7914804 and RG by a grant from Brazil, CNPq 1112.1003/77.

References

1. ALLMAN, J. M. Reconstructing the evolution of the brain in primates through the use of comparative neurophysiological and neuroanatomical data. In: *Primate Brain Evolution Methods and Concepts.*, edited by E. Armstrong. New York: Plenum Press, 1981.

2. ALLMAN, J. M., AND KAAS, J. H. A representation of the visual field in the caudal third of the middle temporal gyrus of the owl monkey *(Aotus trivirgatus). Brain Res.*, 31: 85–105, 1971.

3. ALLMAN, J. M., KAAS, J. H., AND LANE, R. H. The middle temporal visual area (MT) in the bushbaby, *Galago senegalensis. Brain Res.*, 57: 197–202, 1973.

4. BARBER, G., GATTASS, R., AND GROSS, C. G. Unpublished data.

5. BENDER, D. B. Properties of single neurons in the inferior pulvinar of the rhesus monkey. *Assoc. Res. Vision Opthalmol.*, p. 92, 1976.

6. BENDER, D. B. A double representation of the visual hemifield in the monkey pulvinar. *Assoc. Res. Vision Opthalmol.*, p. 151, 1978.

7. BENEVENTO, L. A., AND FALLON, J. H. The ascending projections of the superior colliculus in the rhesus monkey *(Macaca mulatta). J. Comp. Neurol.*, 160: 339–362, 1975.

8. BENEVENTO, L. A., AND REZAK, M. The cortical projections of the inferior pulvinar and adjacent lateral pulvinar in the rhesus monkey *(Macaca mulatta):* an autoradiographic study. *Brain Res.*, 108: 1–24, 1976.

9. BIZZI, E., AND SCHILLER, P. H. Single unit activity in the frontal eye fields of unanesthetized monkeys during head and eye movements. *Exptl. Brain Res.*, 10: 151–158, 1970.

10. BRODMANN, K. Beiträge zur histologischen Lokalization der Grosshirnrinde. Dritte Mitteilung: die Rindenfelder der neideren Affen. *J. Psychol. Neurol, Leipzig,* 4: 177–226, 1905.

11. BRUCE, C. J., DESIMONE, R., AND GROSS, C. G. Properties of neurons in a visual polysensory area in the superior temporal sulcus of the macaque. *J. Neurophysiol.,* 46: 369–384, 1981.

12. BRUCE, C. J., DESIMONE, R., AND GROSS, C. G. Large visual receptive fields in a polysensory area in the superior temporal sulcus of the macaque. *Soc. Neurosci. Abstr.,* 3: 554, 1977.

13. COWEY, A. Cortical maps and visual perception: The Grindley memorial lecture. *Quart. J. Exptl. Psychol.,* 31: 1–17, 1979.

14. COWEY, A., AND GROSS, C. G. Effects of foveal prestriate and inferotemporal lesions on visual discrimination by rhesus monkeys. *Exptl. Brain Res.,* 11: 128–144, 1970.

15. COWEY, A., AND WEISKRANTZ, L. A comparison of the effects of inferotemporal striate cortex lesions on the visual behaviour of rhesus monkeys. *Quart. J. Exptl. Psychol.,* 19: 246–253, 1967.

16. CRAGG, B. G., AND AINSWORTH, A. The topography of the afferent projections in the circumstriate visual cortex of the monkey studied by the Nauta method. *Vision Res.,* 9: 733–747, 1969.

17. DEKKER, J. J., KIEVIT, J., JACOBSON, S., AND KUYPERS, H. G. J. M. Retrograde axonal transport of horseradish peroxidase in the forebrain of the rat, cat, and rhesus monkey. In: *Golgi Centennial Symposium, Proceedings,* edited by M. Santini. New York: Raven Press, 1975, pp. 201–208.

18. DESIMONE, R., ALBRIGHT, T. D., GROSS, C. G., AND BRUCE, C. J. Responses of inferior temporal neurons to complex visual stimuli. *Soc. Neurosci. Abstr.,* 6: 581, 1980.

19. DESIMONE, R., BRUCE, C., AND GROSS, C. G. Neurons in the superior temporal sulcus of the macaque still respond to visual stimuli after removal of striate cortex. *Soc. Neurosci. Abstr.,* 5: 781, 1979.

20. DESIMONE, R., FLEMING, J., AND GROSS, C. G. Prestriate afferents to inferior temporal cortex: an HRP study. *Brain Res.,* 184: 41–55, 1980.

21. DESIMONE, R., AND GROSS, C. G. Visual areas in the temporal cortex of the macaque. *Brain Res.,* 178: 363–380, 1979.

22. GATTASS, R. AND GROSS, C. G. A visuotopically organized area in the posterior superior temporal sulcus of the macaque. *Assoc. Res. Vision Opthalmol.,* p. 184, 1979.

23. GATTASS, R., AND GROSS, C. G. Visual topography of striate projection zone (MT) in posterior superior temporal sulcus of the macaque. *J. Neurophysiol.,* 46: 621–638, 1981.

24. GATTASS, R., OSWALDO-CRUZ, E., AND SOUSA, A. P. B. Visuotopic organization of the cebus pulvinar: a double representation of the contralateral hemifield. *Brain Res.,* 152: 1–16, 1978.

25. GOLDBERG, M. E., AND BUSHNELL, M. C. Personal communication.

26. GROSS, C. G. Visual functions of inferotemporal cortex. In: *Handbook of Sensory Physiology,* edited by R. JUNG. Berlin: Springer, VII/3B, 451–482, 1973.

27. GROSS, C. G. Inferotemporal cortex and vision. In: *Progress in Physiological Psychology*, edited by E. STELLAR AND J. M. SPRAGUE. New York: Academic Press, 5: 77–123, 1973.

28. GROSS, C. G. Inferior temporal lesions do not impair discrimination of rotated patterns in monkeys. *J. Comp. Physiol., Psychol.* 92: 1095–1109, 1978.

29. GROSS, C. G., BENDER, D. B., AND GERSTEIN, G. L. Activity of inferior temporal neurons in behaving monkeys. *Neuropsychologia,* 17: 215–229, 1979.

30. GROSS, C. G., BENDER, D. B., AND MISHKIN, M. Contributions of the corpus callosum and the anterior commissure to the visual activation of inferior temporal neurons. *Brain Res.,* 131: 227–239, 1977.

31. GROSS, C. G., BENDER, D. B., AND ROCHA-MIRANDA, C. E. Inferotemporal cortex: a single unit analysis. In: *The Neurosciences: A Third Study Program*, edited by F. O. SCHMITT AND F. G. WORDEN. Cambridge: M. I. T. Press, 1973, pp. 229–238.

32. GROSS, C. G., AND MISHKIN, M. The neural basis of stimulus equivalence across retina translation. In: *Lateralization in the Nervous System*, edited by S. HARNED, R. DOTY, J. JAYNES, L. GOLDBERG, AND G. KRAUTHAMER. New York: Academic Press, 1977, pp. 109–122.

33. GROSS, C. G., ROCHA-MIRANDA, C. E., AND BENDER, D. B. Visual properties of neurons in inferotemporal cortex of the macaque. *J. Neurophysiol.,* 35: 96–111, 1972.

34. HARTJE, W., AND ETTLINGER, G. Reaching in light and dark after unilateral posterior parietal ablations in the monkey. *Cortex,* 9: 344–354, 1973.

35. HECAEN, H., AND ALBERT, M. L. *Human Neuropsychology*. New York: Wiley, 1978.

36. HUBEL, D. H., AND WIESEL, T. N. Stereoscopic vision in macaque monkey. *Nature,* 225: 41–42, 1970.

37. HUMPHREY, N. K. Vision in a monkey without striate cortex: a case study. *Perception,* 241–255, 1974.

38. INGALLS, V. A., BRUCE, C. J., DESIMONE, R., AND GROSS, C. G. Unpublished data.

39. JONES, E. G., AND BURTON, H. Areal differences in the laminar distribution of thalamic afferents in cortical fields of the insular, parietal and temporal regions of primates. *J. Comp. Neurol.,* 168: 197–247, 1976.

40. JONES, E. G., AND POWELL, T. P. S. An anatomical study of converging sensory pathways within the cerebral cortex of the monkey. *Brain ,* 93: 793–820, 1970.

41. KUYPERS, H. G., SZWARCBART, M. K., MISHKIN, M., AND ROSVOLD, H. E. Occipito-temporal cortico-cortical connections in the rhesus monkey. *Exptl. Neurol.,* 11: 245–262, 1965.

42. LAMOTTE, R. H., AND ACUNA, C. Defects in accuracy of reaching after removal of posterior parietal cortex in monkeys. *Brain Res.,* 139: 309–326, 1978.

43. LATTO, R., AND COWEY, A. Fixation changes after frontal eyefield lesions in monkeys. *Brain Res.,* 30: 25–36, 1971.

44. LATTO. R., AND COWEY. A. Visual field defects after frontal eyefield lesions in monkeys. *Brain Res.*, 30: 1–24, 1971.
45. MANNING. F. J., AND MISHKIN. M. Further evidence on dissociation of visual deficits following partial inferior temporal lesions in monkeys. *Soc. Neurosci. Abstr.*, 2: 1126, 1976.
46. MAUNSELL. J. H. R., BIXBY. J. L., AND VAN ESSEN. D. C. The middle temporal area (MT) in the macaque: architecture, functional properties and topographic organization. *Soc. Neurosci. Abstr.* 5: 796, 1979.
47. MESULAM. M.-M., VAN HOESEN. G. W., PANDYA. D. N., AND GESCHWIND. N. Limbic and sensory connections of the inferior parietal lobule (area PG) in the rhesus monkey: a study with a new method for horseradish peroxidase histochemistry. *Brain Res.*, 136: 393– 414, 1977.
48. MISHKIN. M., AND OUBRE. J. L. Dissociation of deficits on visual memory tasks after inferior temporal and amygdala lesions in monkeys. *Soc. Neurosci. Abstr.*, 2: 1127, 1976.
49. MOHLER. C. W., GOLDBERG. M. E., AND WURTZ. R. H. Visual receptive fields of frontal eye field neurons. *Brain Res.*, 61: 385–389, 1973.
50. MOHLER. C. W., AND WURTZ. R. H. Role of striate cortex and superior colliculus in visual guidance of saccadic eye movements in monkeys. *J. Neurophysiol.*, 40: 74–94, 1977.
51. MOUNTCASTLE. V. B., LYNCH. J. C., GEORGOPOULOS. A., SAKATA. H., AND ACUNA. C. Posterior parietal association cortex of the monkey: command functions for operations within extra-personal space. *J. Neurophysiol.*, 38: 871–908, 1975.
52. MYERS. R. E. Organization of visual pathways. In: *Functions of the Corpus Callosum*, edited by E. G. ETTLINGER. A. V. S. DE REUCH. AND R. POTTER. London: Churchill, 1965, pp. 133–143.
53. PANDYA. D. N., KAROL. E. A., AND HEILBRON. D. The topographical distribution of interhemispheric projections in the corpus callosum of the rhesus monkey. *Brain Res.*, 32: 31–43, 1971.
54. PETRIDES. M., AND IVERSEN. S. D. The effect of selective anterior and posterior association cortex lesions in the monkey on performance of a visual-auditory compound discrimination test. *Neuropsychologia*, 16: 527–537, 1978.
55. PETRIDES. M., AND IVERSON. S. D. Restricted posterior parietal lesions in the rhesus monkey and performance on visuospatial tasks. *Brain Res.*, 161: 63–77, 1979.
56. PIGAREV. I. N., RIZZOLATTI. G., AND SCANDOLARA. C. Neurons responding to visual stimuli in the frontal lobe of macaque monkeys. *Neurosci. Letters*, 12: 207–212, 1979.
57. ROBINSON. D. L., GOLDBERG. M. E., AND STANTON. G. B. Parietal association cortex in the primate: sensory mechanisms and behavioral modulations. *J. Neurophysiol.*, 41: 910–932, 1978.
58. ROCHA-MIRANDA. C. E., BENDER. D. B., GROSS. C. G., AND MISHKIN. M. Visual activation of neurons in inferotemporal cortex depends on striate cortex and forebrain commissures. *J. Neurophysiol.*, 38: 475–491, 1975.

59. ROSENKILDE, C. E. Functional heterogeneity of the prefrontal cortex in the monkey: a review. *Behav. Neural Biol.*, 25: 301–345, 1979.

60. SCHILLER, P. H., AND MALPELI, J. G. The effect of striate cortex cooling on area 18 cells in the monkey. *Brain Res.*, 126: 366–369, 1977.

61. SEACORD, L., GROSS, C. G., AND MISHKIN, M. Role of inferior temporal cortex in interhemispheric transfer. *Brain Res.*, 167: 259–272, 1979.

62. SELTZER, B., AND PANDYA, D. N. Afferent cortical connections and architectonics of the superior temporal sulcus and surrounding cortex in the rhesus monkey. *Brain Res.*, 149: 11–24, 1978.

63. SPATZ, W. B., TIGGES, J., AND TIGGES, M. Subcortical projections, cortical associations, and some intrinsic interlaminar connections of the striate cortex in the squirrel monkey *(Saimiri). J. Comp. Neurol.*, 140: 155–174, 1970.

64. UNGERLEIDER, L. G., AND MISHKIN, M. Personal communication.

65. UNGERLEIDER, L. G., AND MISHKIN, M. The striate projection zone in the superior temporal sulcus of *Macaca mulatta:* location and topographic organization. *J. Comp. Neurol.*, 188: 347–366, 1979.

66. VAN ESSEN, D. C. Visual areas of the mammalian cerebral cortex. *Ann. Rev. Neurosci.*, 2: 227–263, 1979.

67. VAN ESSEN, D. C., MAUNSELL, J. H. R., AND BIXBY, J. L. The organization of extrastriate visual areas in the macaque monkey. In: *Cortical Sensory Organization*, edited by C. N. WOOLSEY. Clifton, New Jersey: Humana Press, 1981, vol. 2, chapter 6.

68. VON BONIN, G., AND BAILEY, P. *The Neocortex of Macaca Mulatta*, Univ. of Illinois press, Urbana, Illinois, 1947.

69. WEISKRANTZ, L., COWEY, A., AND PASSINGHAM, C. Spatial responses to brief stimuli by monkeys with striate cortex ablations. *Brain*, 100: 655–670, 1977.

70. WELLER, R. E., AND KAAS, J. M. Connections of striate cortex with the posterior bank of the superior temporal sulcus in macaque monkeys. *Soc. Neurosci. Abstr.*, 4: 2089, 1978.

71. WHITLOCK, D. G., AND NAUTA, W. J. H. Subcortical projections from the temporal neocortex in *Macaca mulatta. J. Comp. Neurol.*, 106: 183–212, 1956.

72. WILSON, M., KEYS, AND JOHNSTON, T. D. Middle temporal cortical visual area and visuospatial function in *Galago sensgalensis. J. Comp. Physiol. Psychol.*, 93: 247–259, 1979.

73. WURTZ, R. H., AND MOHLER, C. W. Enhancement of visual response in monkey striate cortex and frontal eye fields. *J. Neurophysiol.*, 39: 766–772, 1976.

74. ZEKI, S. M. Representation of central visual fields in prestriate cortex of the monkey. *Brain Res.*, 14: 271–291, 1969.

75. ZEKI, S. M. Cortical projections from two prestriate areas in the monkey. *Brain Res.*, 34: 19–35, 1971.

76. ZEKI, S. M. Comparison of the cortical degeneration in the visual regions of the temporal lobe of the monkey following section of the ante-

rior commissure and the splenium. *J. Comp. Neurol.*, 148: 167–176, 1973.

77. ZEKI, S. M. Functional organization of a visual area in the posterior bank of the superior temporal sulcus of the rhesus monkey. *J. Physiol., London*, 236: 549–573, 1974.

78. ZEKI, S. M. Cells responding to changing image size and disparity in the cortex of the rhesus monkey. *J. Physiol., London*, 242: 827–841, 1974.

79. ZEKI, S. M. The projections to the superior temporal sulcus from areas 17 and 18 in the rhesus monkey. *Proc. Roy. Soc London, B.* 193: 199–207, 1976.

80. ZEKI, S. M. Colour coding in the superior temporal sulcus of rhesus monkey visual cortex. *Proc. Roy. Soc., London, B.* 197: 195–223, 1977.

81. ZEKI, S. M. Uniformity and diversity of structure and function in rhesus monkey prestriate visual cortex. *J. Physiol., London*, 277: 273–290, 1978.

82. ZEKI, S. M. Functional specialization in the visual cortex of rhesus monkey. *Nature*, 274: 423–428, 1978.

Index